STARTER PACKS:
A STRATEGY TO FIGHT HUNGER IN DEVELOPING COUNTRIES?

Lessons from the Malawi Experience 1998–2003

Starter Packs: A Strategy to Fight Hunger in Developing Countries?
Lessons from the Malawi Experience 1998–2003

Edited by

Sarah Levy

Calibre Consultants
Reading
UK

CABI Publishing

CABI Publishing is a division of CAB International

CABI Publishing
CAB International
Wallingford
Oxfordshire OX10 8DE
UK

CABI Publishing
875 Massachusetts Avenue
7th Floor
Cambridge, MA 02139
USA

Tel: +44 (0)1491 832111
Fax: +44 (0)1491 833508
E-mail: cabi@cabi.org
Website: www.cabi-publishing.org

Tel: +1 617 395 4056
Fax: +1 617 354 6875
E-mail: cabi-nao@cabi.org

© CAB International 2005. All rights reserved. No part of this publication may be reproduced in any form or by any means, electronically, mechanically, by photocopying, recording or otherwise, without the prior permission of the copyright owners.

A catalogue record for this book is available from the British Library, London, UK.

A catalogue record for this book is available from the Library of Congress, Washington, DC, USA.

ISBN 0 85199 008 8

Typeset by SPI Publisher Services, Pondicherry, India.
Printed and bound in the UK by Cromwell Press, Trowbridge, UK.

Contents

Map of Malawi	viii
Photograph of Starter Pack Conference Delegates	ix
General Starter Pack Photographs	x
Acknowledgements	xiii
Author Biographical Sketches	xv
List of Acronyms and Glossary	xix

Introduction   1
Sarah Levy

PART 1: ORIGINS AND MANAGEMENT OF STARTER PACK

1. **The Origin and Concept of the Starter Pack**   15
   Malcolm Blackie and Charles K. Mann

2. **The Players and the Policy Issues**   29
   Harry Potter

3. **The Logistics and Costs of Implementation**   41
   Charles Clark

4. **Pack Distribution and the Role of Vouchers**   55
   Anthony Cullen and Max Lawson

PART 2: METHODOLOGY OF THE EVALUATION PROGRAMME

5   Design of the Evaluation Programme   69
Ian M. Wilson

6   Experience and Innovation: How the Research Methods Evolved   77
Carlos Barahona

7   Lessons on Management of Large-scale Research Programmes   93
Sarah Levy

PART 3: LESSONS FROM STARTER PACK

8   Production, Prices and Food Security: How Starter Pack Works   103
Sarah Levy

9   The Farmer's Perspective – Values, Incentives and Constraints   117
Jan Kees van Donge

10   Do Free Inputs Crowd Out the Private Sector in Agricultural Input Markets?   129
Clement Nyirongo

11   Practical and Policy Dilemmas of Targeting Free Inputs   141
Blessings Chinsinga

12   Starter Pack and Sustainable Agriculture   155
Carlos Barahona and Elizabeth Cromwell

13   The Challenges of Agricultural Extension   175
Chris Garforth

14   Why Free Inputs Failed in the Winter Season   193
Hiester Gondwe

15   Financing and Macro-economic Impact: How Does Starter Pack Compare?   203
Sarah Levy

PART 4: BY SPECIAL INVITATION

16 Poverty, AIDS and Food Crisis   219
    Anne C. Conroy

17 Food Security Policies and Starter Pack: a Challenge
    for Donors?   229
    Jane Harrigan

18 Feeding Malawi from Neighbouring Countries   247
    Martin J. Whiteside

19 Starter Pack in Rural Development Strategies   261
    Andrew Dorward and Jonathan Kydd

Conclusion   279
Sarah Levy

Index   289

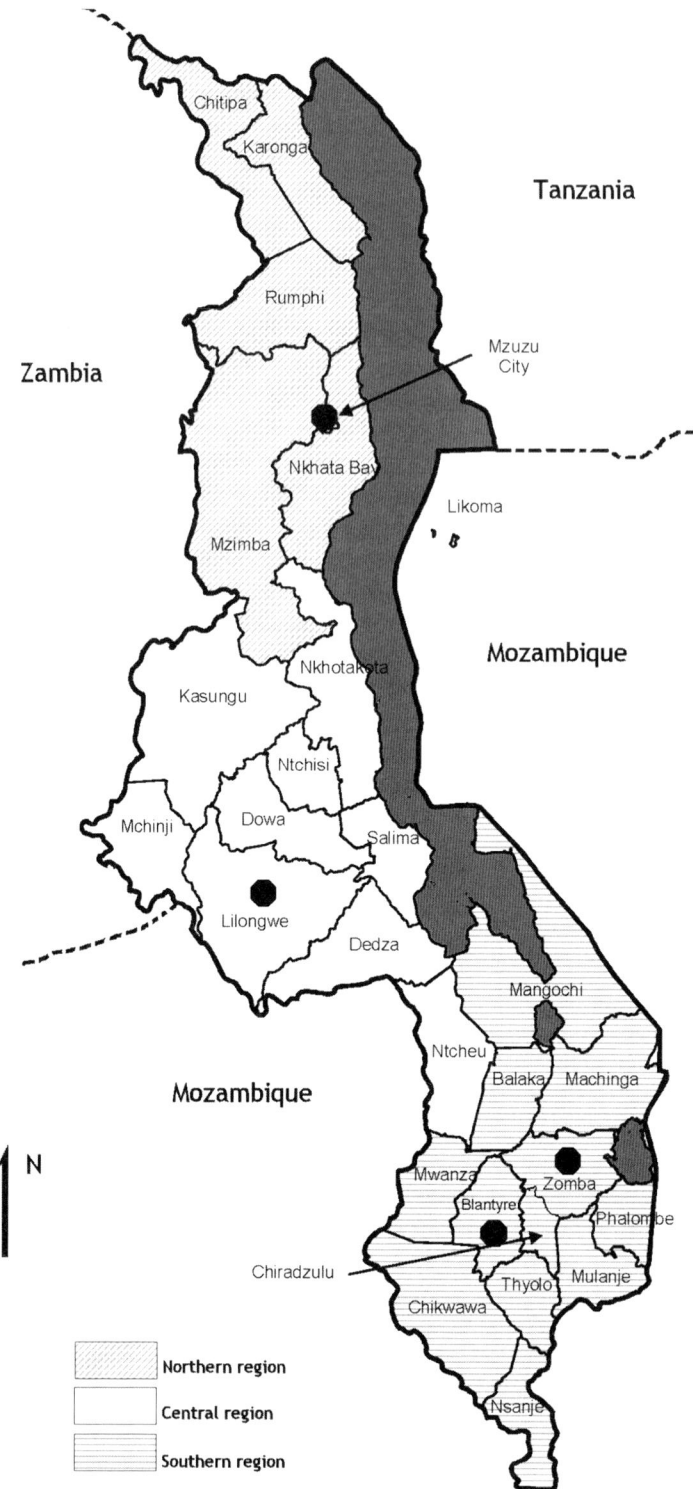

Map of Malawi. (Adapted from a map provided by Todd Benson at IFPRI).

Photograph of Starter Pack Conference Delegates at the Rockefeller Foundation's Bellagio Study and Conference Center, October 2004.

A Starter Pack

Government officials handing out packs at official opening ceremony

Empty warehouse

Warehouse clerks and workers

Full warehouse

Loading a truck at the warehouse

General Starter Pack Photographs

Beneficiary receiving pack at distribution centre

Carrying packs back to the village

Farmer in his Starter Pack plot

Village discussions 1999/2000 Starter Pack evaluation (Kwilindi village)

Farmer in her Starter Pack plot

Village discussions 1999/2000 Starter Pack evaluation (Demula village)

General Starter Pack Photographs

Acknowledgements

The Editor would like to thank Harry Potter (DFID), who commissioned this book and most of the research that went into it; the chapter authors, many of whom were also consultants for the Starter Pack/TIP M&E programme; and the other consultants and researchers who took part in the M&E programme: Frederick Msiska, Humphrey Mdyetseni, Frank Kamanga, Christopher Dzimadzi, Peter Wingfield Digby, Francis Nyirenda, MacNewman Msowoya, Lucy Binauli, Pickford Sibale, Ishmael Gondwe, G. Mazunda, M. Khwepeya, Readwell Msopole, Esthery Kunkwenzu, McLloyd Polepole, Lucy Chipeta, Rowland Chirwa, Patrick Kambewa, Richard Mwanza, Francis Lwanda, Bright Sibale, Stuart Ligomeka, Rodwell Chinguwo, Sidon Konyani, Martin Palamuleni, A. Chirembo, Alex Saka, Victor Lungu, Regson Chaweza, Mufunanji Magalasi, Lawrence Mpekansambo, Boniface Dulani, Happy Kayuni, Mackenzie Chivwaile, William Kasapila, Prince Kapondamgaga, Overtoun Mgemezulu, Noel Sangore and Elarton Thawani.

The Editor is also grateful to the hundreds of research assistants, field supervisors, enumerators and data entry clerks who worked for the M&E programme between 1999 and 2003; and, above all, to the thousands of smallholder farmers of Malawi, who spent time discussing the issues and being interviewed for the surveys.

The Editor would also like to thank the following people: Cecilia Cruz, Alan Whitworth, Lindsay Mangham, Mark Davies, Jimmy Kawaye, Harriet Menter, Margaret Gaynor, Chigomezgo Mtegha, Comfort Khembo, Catherine Hara, Angel Msukwa, Temwa Chirambo and Edna Phiri (DFID, Malawi); Donal Brown, Sharon Harvey, Graham McDonald, Anne Stanley and Fiona McLachlan (DFID, UK); John Hansell and Joanne Manda (DFID, Zimbabwe); Aleke Banda, Ellard Malindi, Andrina Mchiela, Charles Mataya, Alex Namaona, Zakeyo Kamanga, Dan Kamputa, Stanley Chimhonda, Grace Malindi, Bright Nsendema, Dan Yona, C. Chowa, G. Ghavula, Ken Ndindi and Agnes Kazembe (MoA, Malawi); George Ndovi, Francis Kathewera,

Chimwemwe Phiri, Fainess Mambya, Liners Hamuza, Zimveka Kazembe and Darlington Matthews (Starter Pack/TIP Logistics Unit, Malawi); Patrick Kabambe (Office of the President and Cabinet, Malawi); James Phiri, Chauncy Simwaka and George Zimalirana (MoF, Malawi); Milton Kutengule (MEP&D, Malawi); Wilson Banda and Grant Kabango (Reserve Bank of Malawi); Dan Saukila (NFRA, Malawi); Sam Chimwaza, Evance Chapasuka, Joan Chalira and Anne Botha (FEWS, Malawi); B. Ng'oma (Ministry of Labour, Malawi); Charles Machinjili (NSO, Malawi); Gerard van Dijk, Lola Castro, Masozi Kachale and Abdelgadir Hamid (WFP, Malawi); Lawrence Rubey (USAID, Malawi); Herschel Weeks (IFDC/AIMS); Paul Ginies, Isabelle Le Normand, Elizabeth Sibale and Duncan Samikwa (EU, Malawi); Tori Hoven (Royal Norwegian Embassy, Malawi); Nick Osborne and Boster Sibande (CARE, Malawi); Eleanor Allen and Lorna Turner (SSC, UK); Toby Kaima (DFID Guesthouse, Malawi); Symon Ngulama (Andrews Car Hire, Malawi); Montgomery Thunde (independent graphic artist, Malawi); Lindsay McConaghy (Ulendo Safaris, Malawi); Nicholas Freeland (MTLconsult, France); Mick Foster (Mick Foster Economics, UK); Michael Pickstock (Wren Media, UK); Steve Wiggins and John Farrington (Overseas Development Institute, UK); Neil Fantom (NSO, Malawi, later World Bank); and Todd Benson (National Economic Council, Malawi, later IFPRI).

Finally, the Editor expresses her sincere thanks to The Rockefeller Foundation for hosting and providing financial support for a conference on the theme of this book at the Foundation's Bellagio Study and Conference Center, Como, Italy, in October 2004. In particular, the Editor would like to thank the following people at The Rockefeller Foundation: John Lynam and Wanjiku Kiragu (Nairobi office); Susan E. Garfield and C. Jocelyn Peña (New York office); and Gianna Celli and Nadia Gilardoni (Bellagio office).

Author Biographical Sketches

Carlos Barahona is a Senior Statistician at the Statistical Services Centre (SSC), University of Reading (UK), where his main duties involve consultancy and training. He has a BSc in agriculture and an MSc in biometry. His work has taken him to Malawi, Zambia, Kenya, Uganda, Ethiopia, Jamaica, Bolivia, Honduras and Guatemala. His interests include the use of statistics in agriculture, planning and supporting optimal research processes in developing countries, and the integration of statistical and participatory methods. He was the technical manager for the Starter Pack Monitoring and Evaluation (M&E) programme in Malawi.

Malcolm Blackie, a Zimbabwean, worked outside Africa during much of the period following his country's unilateral declaration of independence in 1965. Invited to return to Zimbabwe in 1980, he spent most of the next decade building the Faculty of Agriculture at the University of Zimbabwe. In 1988, he joined the Rockefeller Foundation and was responsible for the establishment of its Southern Africa Agricultural Sciences Programme. He was a member of the team that developed the original Starter Pack proposal. Malcolm Blackie left the Foundation in 1999 and now works as an author and consultant on smallholder agricultural development and natural resource issues.

Blessings Chinsinga is a Senior Lecturer in poverty alleviation, institutional analysis and decentralization at the Department of Political and Administrative Studies, Chancellor College, University of Malawi. He was team leader for several of the 'qualitative' studies looking at food security, safety nets and community poverty targeting during the Starter Pack evaluations. He is currently studying for a PhD at the University of Bonn, Germany.

Charles Clark, a surveyor by profession, worked for 30 years in the Malawi Ministry of Works, during which time he was responsible for the development of the new capital city of Lilongwe, concluding his career in the Ministry as Principal Secretary. He then joined the World Food Programme (WFP), heading the emergency unit during the 1992 drought in southern Africa. Subsequently, he was responsible for the Mozambique refugee repatriation programme. Since leaving WFP, he has functioned as a consultant for the UK Department for International Development (DFID) in various emergency interventions. Since 1998, he has managed the Logistics Units for the free inputs programmes (Starter Pack and its successors).

Anne Conroy has worked in Malawi since 1987. She first went to the country as a Voluntary Service Overseas (VSO) volunteer in the Ministry of Agriculture (MoA). She did her PhD research on the use of seed and fertilizer by smallholder farmers in Malawi. She worked for the Malawi Government from 1994 to 2004 in the Ministry of Economic Planning and Development (MEP&D) and the Office of the Vice President. She was Special Assistant to the Vice President on Agriculture and Food Security, Health and HIV/AIDS. She was part of the team that designed Starter Pack and later a member of the National Food Crisis Joint Task Force. She is currently working on a publication *Africa's Perfect Storm: Poverty, HIV/AIDS and Hunger in Malawi* with co-authors Justin Malewezi, Jeffrey Sachs and Alan Whiteside.

Elizabeth Cromwell is a Research Fellow in the Rural Policy and Governance Group at the Overseas Development Institute (London), a leading independent think-tank on development policy issues. An agricultural economist by training, she specializes in analysis and policy design relating to rural livelihoods and sustainable agriculture in eastern and southern Africa. In the 1980s she worked as an Evaluation Officer for the Malawi Government. She was subsequently involved with a number of agricultural policy initiatives in the country, including the 1999/2000 Starter Pack evaluation.

Anthony Cullen is a PhD Candidate and Irish Research Council for the Humanities and Social Sciences (IRCHSS) Government of Ireland Scholar at the Irish Centre for Human Rights, National University of Ireland, Galway. In addition to food insecurity in sub-Saharan Africa, his other areas of research interest include the role of international human rights standards in sustainable development, good governance and national implementation of international humanitarian law. He was a member of the Monitoring Component team of the 2000/01 Starter Pack M&E programme.

Andrew Dorward is an agricultural development economist with particular interests in micro-economic and institutional analysis to inform policies promoting poverty-reducing rural growth in developing countries. He is currently a Reader in Agricultural Development Economics in the Department of Agricultural Sciences, Imperial College, London. He has strong links with Malawi and has worked on short- and long-term research, training and development assignments in Africa, Asia and Latin America.

Author Biographical Sketches

Chris Garforth is Professor of Agricultural Extension and Rural Development at the University of Reading's School of Agriculture, Policy and Development. He did his PhD on the relationship between land tenure and agricultural land use in Nigeria. He first worked in extension research in Botswana in the late 1970s and has subsequently undertaken research projects and consultancy assignments in the fields of communication, extension and knowledge transfer in more than 40 countries in Europe, Africa and Asia.

Hiester Gondwe holds an MSc in Agricultural Economics. He works for the Malawi MoA where he is currently Head of the Monitoring and Evaluation Section in the Planning Department. He was a member of the Starter Pack evaluation team from 1999 to 2003, taking a leading role in the main season surveys and the 2003 winter season survey.

Jane Harrigan is a Senior Lecturer in Development Economics in the School of Environment and Development at the University of Manchester. Prior to taking up her post in Manchester she worked as an economist in the MoA, Malawi, and has published extensively on economic policy in Malawi, including the book *From Dictatorship to Democracy*. She is also co-author of the book *Aid and Power: The World Bank and Policy Based Lending*. Her current research interests include IMF and World Bank programmes in the Middle East and North Africa.

Jonathan Kydd is Professor of Agricultural Development Economics at Imperial College, London. From 1975 to 1983 he was on the staff of Chancellor College, University of Malawi. Since leaving Chancellor College, he has continued to publish on development issues affecting Malawi.

Max Lawson is a Policy Adviser on Social Policy and Governance in the Policy Department of Oxfam (UK). Before joining Oxfam, he worked for two and a half years in Malawi. He co-ordinated the Monitoring Component of the Starter Pack M&E programme in 2000/01. He has a BA in politics and philosophy and an MA in rural development.

Sarah Levy began her career as an economist teaching at the University of El Salvador. On returning to the UK in 1994, she worked as a macro-economic analyst/editor for the Economist Intelligence Unit (London) and as head of the Latin America desk at Oxford Analytica (Oxford). In 1998, she set up Calibre Consultants (UK), which focuses on data collection, economic analysis and management of research in developing countries. She was part of the team that managed the M&E programme for Starter Pack from 1999 to 2004. She is currently a Visiting Research Fellow at the School of Agriculture, Policy and Development at the University of Reading, where she lectures in Development Finance.

Charles Mann served for nearly 10 years as food security adviser to the government of Malawi and was a member of the team that designed the

Starter Pack programme. He joined the Harvard Institute for International Development in 1985 and is now a retired fellow of the Harvard Center for International Development. Before moving to Harvard, he was Chief of the Economic Analysis Staff of the United States Agency for International Development (USAID) mission in Turkey and later Associate Director for Agricultural and Social Sciences at the Rockefeller Foundation. In 2003 he founded the International Development Communications Workshop to produce video documentation of development issues.

Clement Nyirongo is Deputy Chief Economist in the Development Division of the MEP&D in Malawi. He was the team leader for several of the 'food production and security' studies during the Starter Pack and Targeted Inputs Programme (TIP) evaluations from 2000 to 2003.

Harry Potter was a Livelihoods Adviser at DFID until he retired in August 2004. He has more than 35 years of experience in rural development. He was based in Malawi from 1996 to 2004 and is an expert on the country's agriculture sector and its agriculture and food security policies. He played a leading role in donor–government negotiations on the design and implementation of the Starter Pack programme from its inception in 1998, and was the driving force behind donor financing for the programme.

Jan Kees van Donge is a Senior Lecturer in Public Policy and Development Management at the Institute of Social Studies, the Hague (Netherlands). He has been a member of staff of the Universities of Zambia, Dar es Salaam and Malawi (Chancellor College, Zomba) over a period of 20 years. In the Netherlands, he has worked for the Agricultural University of Wageningen, where he also did his PhD.

Martin J. Whiteside is a UK-based independent rural development consultant. He has worked for a variety of Non-governmental Organizations (NGOs) and for the United Nations Environment Programme. He has a degree in agriculture and forest sciences and 25 years of experience in programme evaluation, rural livelihood programme design, the use of participatory methodologies and sustainable natural resource management. He has conducted several studies on the informal agricultural trade and other livelihood interactions between Malawi and neighbouring countries, particularly northern Mozambique.

Ian Wilson is a Senior Lecturer in Applied Statistics at the University of Reading and Special Adviser to the SSC, having been its Director from 1983 to 2001. He has an interest in practical aspects of methodological issues broadly related to statistics. His consulting practice involves monitoring and evaluation, surveys and sampling issues in relation to development. He has worked widely in Africa and Asia for 30 years in statistical education and has advised various food security, natural resources and social projects. He designed the approach for the Starter Pack M&E programme.

List of Acronyms and Glossary

US$	US dollar
ABSA	A South African commercial bank
ACB	Agricultural Communication Branch (MoA Extension Department)
ADMARC	Agricultural Development and Marketing Corporation
AIDS	Acquired Immune Deficiency Syndrome
AIMS	Agricultural Input Markets Development Project (supported by USAID)
ALDSAP	Agricultural and Livestock Development Strategy and Action Plan
APIP	Agricultural Productivity Investment Programme
ASP	Agricultural Services Programme (World Bank)
ATC	Authority To Collect form
ATF	Area Task Force
CARE	An independent international aid agency with members in Europe, Japan, North and South America
CD	Compact Disc
CGE	Computable General Equilibrium (model)
CKP	Charles Kendall and Partners
CMR	Crude Mortality Rate
CNFA	Citizens Network for Foreign Affairs
CTS	Commodity Tracking System
DA	District Assembly (district-level local government body)
DADO	District Agricultural Development Officer
Dambo	Wetland used for winter cultivation; this term is used interchangeably with *dimba* in this book (see Chapter 14)
DC	District Commissioner

DFID	Department for International Development (the official UK aid agency)
Dimba	Wetland used for winter cultivation; this term is used interchangeably with *dambo* in this book (see Chapter 14)
DRC	Democratic Republic of Congo
DRIP	Drought Recovery Inputs Programme
DTF	District Task Force
DWT	Direct Welfare Transfer
EC	European Commission
EGS	Employment Generation Scheme
EPA	Extension Planning Area (an administrative unit of the MoA comprising a number of sections staffed by extension workers, formerly known as FAs)
ETIP	Extended Targeted Inputs Programme
EU	European Union
FA	Field Assistant (extension worker)
FAO	Food and Agriculture Organization (of the UN)
FEWS	Famine Early Warning System
FGD	Focus Group Discussion
FY	Financial Year
Ganyu	Agricultural piecework (see Chapter 9)
GDP	Gross Domestic Product
GoM	Government of Malawi
GPS	Global Positioning System
GTIS	Ground Truth Investigation Study
h	Hour(s)
HIPC	Highly Indebted Poor Country
HIV	Human Immunodeficiency Virus
ICT	Information and Communication Technology
IFAD	International Fund for Agricultural Development of the UN
IFDC	International Center for Soil Fertility and Agricultural Development
IFPRI	International Food Policy Research Institute
IHS	Integrated Household Survey
IMF	International Monetary Fund
IRCHSS	Irish Research Council for the Humanities and Social Sciences
IT	Information Technology
LDC	Less Developed Country
List A	The early round of ETIP pack distribution in 2002/03
List B	The late round of ETIP pack distribution in 2002/03
LU	Logistics Unit (the unit co-ordinating implementation of Starter Pack, TIP and ETIP)
M&E	Monitoring and Evaluation
MAFSP	Multi Annual Food Security Programme of the EC
MASAF	Malawi Social Action Fund
MASIP	Malawi Agricultural Sector Investment Plan, later Process

MDHS	Malawi Demographic and Health Survey
MEP&D	Ministry of Economic Planning and Development
MH17	Malawi Hybrid 17
MH18	Malawi Hybrid 18
MK	Malawi Kwacha (local currency)
MoA	Ministry of Agriculture and Irrigation, later Ministry of Agriculture, Irrigation and Food Security
MoF	Ministry of Finance
MPTF	Maize Productivity Task Force
MRFC	Malawi Rural Finance Company
MTEF	Medium Term Expenditure Framework
NASFAM	National Smallholder Farmers Association of Malawi
NASSPA	National Smallholder Seed Producers' Association
NFRA	National Food Reserve Agency
NGO	Non-governmental Organization
NIE	New Institutional Economics
NORAD	The official aid agency of Norway
NSCM	National Seed Company of Malawi
Nsima	A porridge normally made of maize, which is Malawi's staple food
NSNS	National Safety Net Strategy
NSO	National Statistical Office (Malawi)
OFD	On-farm Demonstration (plot of land set up by extension workers on a farmer's land for demonstration purposes)
OPV	Open Pollinated Variety (maize seed)
POD	Proof of Delivery form
PPE	Pro-poor Expenditure
PRA	Participatory Rural Appraisal
PRSP	Poverty Reduction Strategy Paper
PWP	Public Works Programme
RBM	Reserve Bank of Malawi (the Central Bank)
RDP	Rural Development Project (an administrative unit of the MoA, comprising a number of EPAs)
RDR	Recommended Daily Requirement of calories
SACA	Smallholder Agricultural Credit Administration
Sasakawa Global 2000	An African agricultural technology project promoting high-yielding cereal varieties supported by the Sasakawa Africa Association (Japan) and the Carter Center (US)
SDR	Special Drawing Rights (an artificial currency used by the IMF)
SFFRFM	Smallholder Farmers Fertilizer Revolving Fund Mechanism
SGR	Strategic Grain Reserve
SPLIFA	Sustaining Productive Livelihoods through Inputs For Assets (a project supported by USAID and DFID)
SPLU	Starter Pack Logistics Unit
SSC	Statistical Services Centre of The University of Reading
t	Tonne(s)

T&V	Training and Visit (agricultural extension system)
TA	Traditional Authority (a unit of local government below a DA)
TB	Tuberculosis
TIP	Targeted Inputs Programme
TIPLU	Targeted Inputs Programme Logistics Unit
UK	United Kingdom
UN	United Nations
UNFPA	United Nations Population Fund
UNHR	United Nations Humanitarian Response
UNICEF	United Nations Children's Fund
US	United States
USAID	United States Agency for International Development (the official aid agency of the United States)
VSO	Voluntary Service Overseas
VTF	Village Task Force
Ward	The smallest unit of local government, below a TA
WFP	World Food Programme
WHO	World Health Organization

Introduction

SARAH LEVY

What this Book is About

This book is about the Starter Pack and Targeted Inputs Programmes, which were implemented in Malawi from the 1998/99 agricultural season to the 2003/04 season. The original idea of Starter Pack was to give a tiny bag of free agricultural inputs – fertilizer and seed – to every smallholder farmer in Malawi. In its first two years, Starter Pack was successful in achieving national food security. The scaling down of the programme in the 2000/01 season was a major contributor to the food crisis which hit Malawi in early 2002.

Despite repeated interventions by governments, donors and non-governmental organizations (NGOs) in recent years, food insecurity in developing countries continues to be a major problem, and they are forced to rely on food aid again and again. For once, with Starter Pack, we have a success story about how hunger can be tackled by pre-emptive measures, significantly reducing the risk of food crises. This book examines how this was done in Malawi, and asks whether Starter Pack works only in the conditions of that country, or whether it could be replicated elsewhere.

Poverty and Food Insecurity in Rural Malawi

However you measure it, poverty in rural Malawi is extreme and widespread. The most recent Integrated Household Survey in 1997/98 (National Economic Council, 2000) found that 61% of the rural population was below the poverty line (see Chapter 16). With a rural population estimated at 11.52 million in 2000 (see Chapter 6), this means that some seven million people in rural areas are living below the poverty line.

Food is the most basic of all basic needs for poor people, and is closely associated with their definition of poverty. In the 2001/02 season, our

©CAB International 2005. *Starter Packs: a Strategy to Fight Hunger in Developing Countries?* (ed. S. Levy)

research, using participatory methods, estimated that around one-third of smallholder farmers were extremely food insecure (see Chapter 11). There was also extreme seasonal variation according to our survey data (see Chapter 6): extreme food insecurity affected less than 5% of rural households in May–November 2001, rising to nearly 50% of rural households at the peak of the 2002 'hungry period' in March, just before the main maize harvest.

The 2001/02 season followed a bad maize harvest in 2001. Smallholder farm households ran out of maize – the main staple food – faster than normal and sharp increases in the price of maize meant that poor people could not afford to buy food (see Chapter 8). However, we should be careful about our use of terminology here: people tend to refer to a 'bad' harvest and a 'normal' year, as if a bad harvest was unusual and things would normally be better. In fact, the main problem in Malawi is one of chronic underproduction of food (see Chapter 1). As we discuss below, in the liberalized, non-interventionist context, 'bad' maize production years are in fact only too 'normal'. The only good years since agricultural liberalization in the mid-1990s have been those when there have been large-scale interventions to boost food production.

Food security strategies

What options do Malawi and other developing countries have for tackling hunger? What should be their strategy for achieving food security? In Malawi in recent years, much time and effort has been devoted to discussing the options, but there is little clarity about which options should be prioritized – in other words, what the government's strategy should be. Figure 1.1 shows the types of intervention that are possible (see Chapter 17).

Figure 1.1 distinguishes between two broad categories of intervention: those designed to achieve national, aggregate food security; and those which focus on food security at household and individual levels. This distinction is important. Some people assume that national food security can be achieved with small-scale interventions aimed at improving household food security, such as crop diversification, improving livelihoods or providing safety nets. However, while such interventions undoubtedly have a place in poverty reduction and food security strategies, they should not be confused with interventions capable of achieving national, aggregate food security. To achieve this goal, it is necessary to have national self-sufficiency in food production, national capacity to import food (enough foreign exchange) or a combination of the two.

Starter Pack is an example of a large-scale intervention of the 'national, aggregate food security through self-sufficiency in production' variety.[1] I would argue that it is a more efficient way of achieving national, aggregate food security than the alternative route of importing food (see Chapter 15), although imports may also have a role to play (see Chapter 18). Thus, my focus is on the production side. However, it is not my intention to argue that Starter Pack is the only food security intervention that is required. Some of the other interventions in Fig. 1.1 may be seen as complementary. Which ones

Introduction

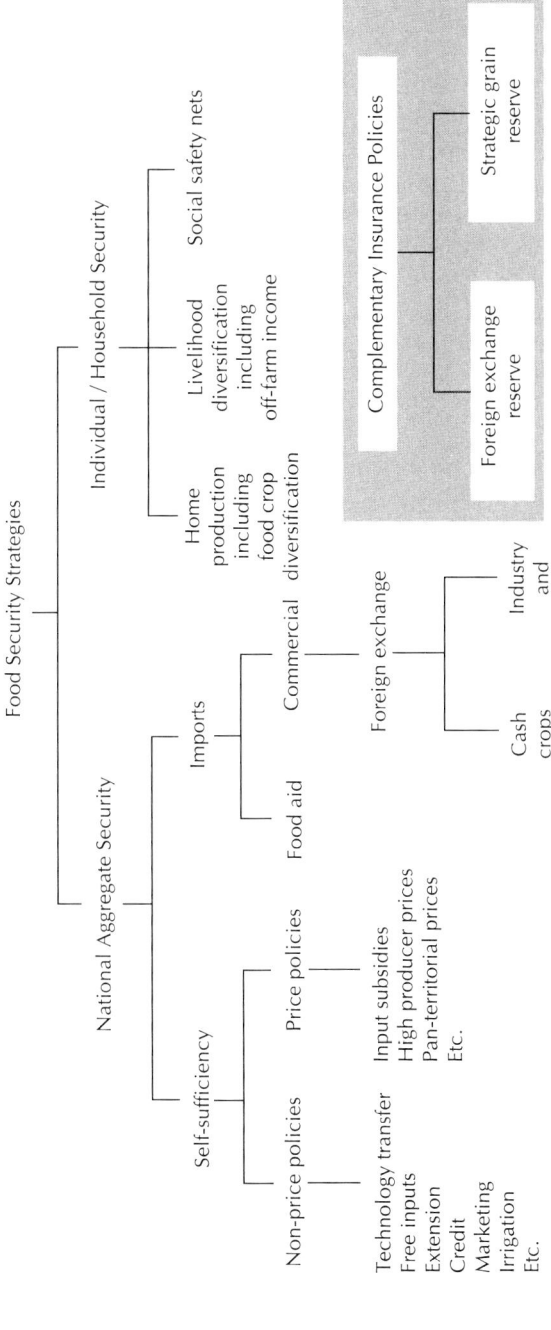

Fig. 1.1. Food security strategies. (Jane Harrigan – presentation to the conference on 'Starter Packs: A Strategy to Fight Hunger in Developing Countries?' at the Rockefeller Foundation's Bellagio Study and Conference Center, October 2004.)

should be included and which should be avoided depends on the broader policy objectives.

The Starter Pack Programme

From 1998/99 to 2003/04, the government of Malawi, with the support of international donors, implemented a free inputs programme for smallholder farmers in the main agricultural season.[2] The programme distributed tiny packs containing roughly 0.1 ha-worth of fertilizer, maize seed and legume seed. Originally, it was known as Starter Pack and it had universal coverage; nearly three million packs were distributed – enough for the whole smallholder population – in 1998/99 and 1999/2000. In the 2000/01 and 2001/02 seasons, however, it was scaled down and efforts were made to target the free inputs to the poorest smallholders. To reflect this change, it became known as the Targeted Inputs Programme (TIP). In 2002/03 and the following year, as a response to a serious food crisis in early 2002, the programme was expanded to near-universal coverage and known as the Extended Targeted Inputs Programme (ETIP).

Starter Pack can be seen as a targeted subsidy for maize production. Even at universal coverage, it only goes to smallholder farmers. Unlike a subsidy on the price of fertilizer, it does not benefit Malawi's estate farmers, which include commercial farms and medium-size 'graduated' smallholders (Mann, 1998). The average beneficiary of Starter Pack is a very small farmer with around 1 ha of cultivated land, little (if any) livestock and no draft power. These farmers are generally poor, and their livelihood options are extremely limited.

In order to have an impact on national, aggregate food security in the current conditions in Malawi, free inputs distributions have to take place on a large scale, reaching most – if not all – smallholders in every part of the country (see Chapter 8). The focus of this book is on this 'universal' or 'near-universal' type of free inputs programme. When discussing the programme in this form – from which we can learn lessons for the future and consider replications in other developing countries – we use the label 'Starter Pack'. References to TIP and ETIP appear only in relation to the specific experiences of the scaled-down and scaled-up versions of the programme.

It can be argued that Starter Pack is not the most appropriate name for the variant of the programme which evolved in Malawi (see Conclusion). However, the debate about names is, in our view, one that should be left to policy makers, who will decide on the best alternative for their particular situation.

Effectiveness and efficiency

The Starter Pack programme was built on 'Best Bet' technologies for maize production, which were the result of five years of agricultural research, including field trials throughout Malawi (see Chapter 1). This research provided a sound basis for the programme. It demonstrated the effectiveness of providing smallholders with small packs containing new varieties of semi-flint hybrid maize, appropriate doses of fertilizer and legumes to improve soil fertility. It was estimated that this technology package would yield 1800 kg of maize per ha, compared with 800 kg for unfertilized maize. Thus, beneficiary farmers would produce an extra 100 kg of maize on average. This proved to be a conservative estimate: farmers were able to produce between 100 and 150 kg of extra maize for each pack of inputs that they received, depending on weather conditions.

At universal coverage (2.8 million packs), Starter Pack contributes between 280,000 and 420,000 t of maize; this is sufficient to overcome chronic food insecurity at national level (see Chapter 8). As part of the research for this book, we carried out a 'least-cost comparison' – similar to cost–benefit analysis but used in cases where benefits are intangible – comparing the annual cost of Starter Pack with those of alternative projects also considered capable of delivering national food security (see Chapter 15). This shows that:

- Starter Pack has a fixed cost of around US$20 million for 2.8 million beneficiaries;
- a general fertilizer subsidy would cost *at least* US$20 million, but costs would probably escalate and would be difficult to predict;
- maize imports would cost US$70 million to US$100 million;
- food aid would cost some US$100 million;
- safety nets in the form of welfare transfers for 30% of rural households would cost around US$107 million.[3]

In addition to being the least-cost solution to the problem of national food security in Malawi, Starter Pack is an efficient solution. Over 80% of programme expenditure in 2002/03 went to beneficiaries in the form of inputs, while the costs of the Ministry of Agriculture's (MoA) semi-autonomous Logistics Unit, which managed the pack distribution, comprised a mere 1% of total expenditure (see Chapter 3). However, this may not be the case in other contexts. Those interested in implementing similar programmes in other countries should carry out careful analysis of the likely costs and benefits of the programme in comparison to the alternatives, and should also examine whether packs can be delivered efficiently.

The end of Starter Pack?

Although free inputs had been distributed before in Malawi and in other countries in the region (see, for example, Munro, 2003, on Zimbabwe), the characteristics of Starter Pack were unique. It was built on proven technologies, and was capable of delivering small amounts of inputs to large numbers of farmers – thus maximizing impact on food security, soil fertility and agricultural diversity while minimizing adverse impact on agricultural inputs markets. Over the six years in which Starter Pack evolved, there was a process of gradual adaptation; but there was also continuity.

In the 2004/05 season, following the election of Bingu wa Mutharika as president of Malawi, the government decided to implement a radically different free inputs programme. The aim was to move gradually towards a general fertilizer subsidy (an election promise) and to involve retailers in pack distribution. In 2004/05, policy makers decided to spend MK4 billion (some US$37 million) on providing two million smallholder farmers with packs containing 25 kg of fertilizer, 5 kg of maize seed and 1 kg of legumes, as well as distributing vouchers for buying fertilizer to 500,000 smallholders. The programme has been highly politicized and is poorly designed (for instance, ignoring retailers' serious supply constraints – see Chapter 4); it has been implemented late; and there is no independent Monitoring and Evaluation (M&E) programme in place to assess its impact.

This book does not attempt to assess the 2004/05 season free inputs distribution. It is to be hoped that it will not be confused with the experience of the previous six years, and – if it proves to be a less cost-effective and efficient alternative to Starter Pack – that decision makers will have the courage to revive Starter Pack in coming years.

Liberalization and Instability

In order to understand the role of Starter Pack in Malawi, it is necessary to know something about the policy changes that have taken place recently. Farmers in the early 2000s work in a very different policy environment from that of the early 1990s. The changes are not unique to Malawi; they are mirrored in other African countries and indeed in other parts of the developing world. The reason that similar changes have occurred in many developing countries is that they are part of a package of stabilization policies and structural adjustment reforms promoted by the IMF and the World Bank.

Two elements of this package have had a particular impact on the food security status of Malawi's smallholder farmers:

1. The agricultural liberalization measures. Until the mid-1990s, smallholder farmers enjoyed subsidies on fertilizer and hybrid maize seed. They could also access cheap credit and sell their produce at supported prices to the Agricultural Development and Marketing Corporation (ADMARC – the state marketing board). By 1993, the smallholder agricultural credit system was facing collapse. Meanwhile, with 'democratization' and the election of President Bakili Muluzi in 1994, agricultural liberalization – begun a few years before under Life President Kamuzu Banda – was accelerated. By 1996, fertilizer and hybrid maize seed subsidies had been removed and agricultural markets had been liberalized. In the late 1990s and early 2000s, smallholders faced rapidly deteriorating terms of trade, volatile markets and scarce credit in place of the previous system of maize production incentives and (artificially) stable markets.

2. A minimal-intervention approach to monetary policy and exchange rate management at the Reserve Bank of Malawi (RBM, the Central Bank), combined with a lack of progress on fiscal reform. The government continues to run serious fiscal deficits which put upwards pressure on inflation and interest rates. Fiscal imbalance (among other things) contributes to rapid depreciation of the local currency, the Malawi Kwacha, against the US dollar. The RBM does not attempt to intervene to slow the depreciation of the Kwacha, partly out of conviction and partly because any such attempt is seen as futile. As a result of this macro-economic instability – which could perhaps better be described as low-level equilibrium stability – farmers find it almost impossible to borrow (the commercial bank lending rate was over 40% in 1999–2003), and they face rapidly rising local currency prices of imported fertilizer (Fig. 1.2), which is essential for growing maize in most parts of Malawi. By the mid-1990s, with soil fertility declining, smallholders were heavily dependent on fertilizer to boost maize yields.

There has been, and will continue to be, much debate about whether this package of policy change and reform is appropriate for developing countries

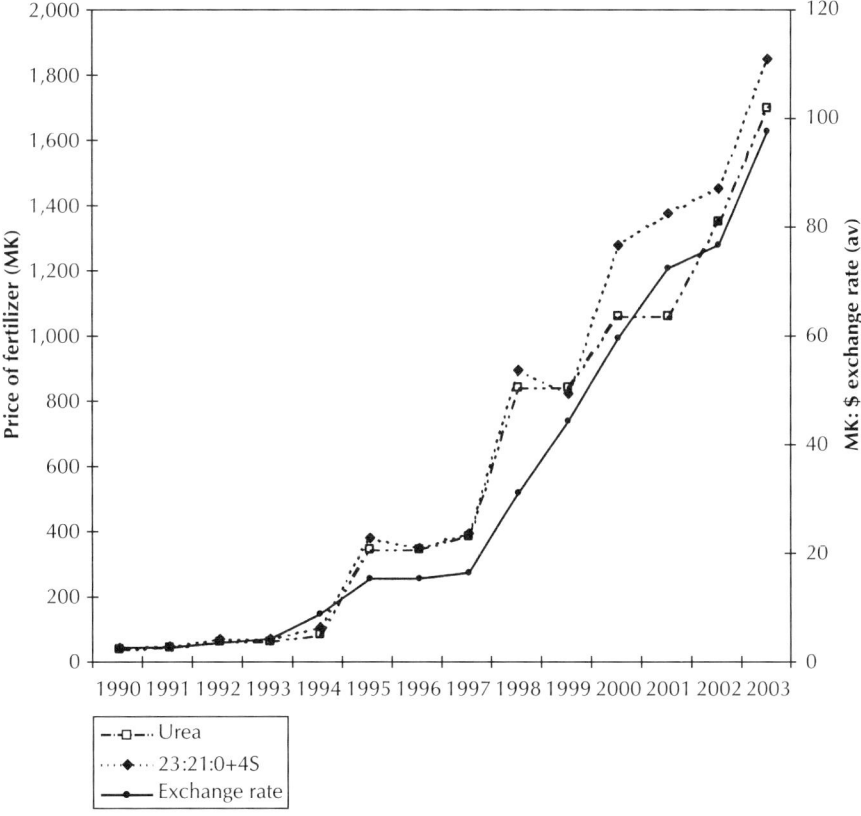

Fig. 1.2. The exchange rate and the price of fertilizer. (IFDC (price of fertilizer) and IMF: International Financial Statistics; Reserve Bank of Malawi (exchange rate)).

like Malawi. Some people argue that the direction of change is generally positive, and it is to be expected that there will be some problems of 'transition' to the free market economy. Others believe that market failure is widespread and therefore a more interventionist approach is needed to promote development. This book does not attempt to address this debate. Our purpose here is simply to point out that conditions have deteriorated for smallholder farmers, particularly with regard to maize production. This may be a transitional phenomena – but if this is the case, it is clearly a long transition process. Therefore, some kind of intervention is needed if serious food insecurity is to be avoided.

Food Crises

When there is a food crisis in Africa, the first reaction of most governments and external observers is to blame the weather. If bad weather is the cause of the crisis, it is logical to conclude that little or nothing could have been done to prevent it and we can wash our hands of responsibility when people go hungry. However, many food crises are in fact the result of poor policies and lack of strategic thinking.

The food crisis that hit Malawi in early 2002 is a case in point. Policy makers were quick to blame bad weather for undermining the 2001 maize harvest, but the 2000/01 rainy season was in fact only moderately bad. More important reasons for the severe – though short-lived – hunger in 2002, following the 2001 harvest, were:

- The changes in smallholder farmers' policy environment such that price incentives for maize production were reduced, few farmers could afford to buy the fertilizer needed to boost maize yields and most lacked access to credit to finance maize production; in particular, farmers reported a sharp increase in local currency prices of fertilizer in the 2000/01 season. This contributed to a maize harvest of only 1.5 million t in 2001, implying a deficit of 500,000 to 600,000 t.
- The Starter Pack programme had been scaled down to become the TIP, with 1.5 million beneficiaries in 2000/01, and had not been replaced by any other large-scale food security programme, reflecting a lack of consensus between policy makers and donors on the importance of a large-scale food security intervention to bridge the country's food gap. The TIP packs were also delivered late – many of them too late to use – in the 2000/01 season.
- The sale of maize from the country's Strategic Grain Reserve in 2001, in a manner that lacked transparency and was followed by speculative marketing of maize.
- The decision by some donors not to react to the low 2001 maize harvest in the mistaken belief that there were other sources of food available to people in rural areas.

There is an important lesson that needs to be learnt from the 2002 food crisis: human beings can cause crises by a combination of actions (or inactions) which may be valuable for their contribution to certain policy goals but which fail to address food security imperatives. If major policy changes are implemented in a poor, food-insecure country like Malawi, it is essential to have a well-thought-out food security strategy in place. In the case of Malawi, without such a strategy, chronic food under-production leads to recurrent food crises. These are costly both in terms of immediate human suffering and because they set back other development initiatives (see Chapters 15 and 16). Policy makers and donors now have as part of their 'toolbox' the option of a programme like Starter Pack that is capable of preventing such regular disasters.

Evidence-based Policy Making

The Starter Pack programme has been researched like no other programme in Malawi in the last decade. The original design was built on solid research into maize production technologies. When implementing the programme, the Logistics Unit kept detailed records and wrote regular reports, providing the basis for the analysis presented in Chapter 3. From the first year of implementation, independent M&E teams were called in to advise on how to assess the impact of the programme. From 1999 to 2003, the M&E programme was managed by a UK-based team from the Statistical Services Centre at The University of Reading and Calibre Consultants, who coordinated a total of 18 research modules ranging from surveys to case studies. The studies were designed and implemented by Malawian and Malawian–European teams, with strong support from the managers, who ensured that they met rigorous quality control and ethical standards. The 1999–2003 M&E archive is available on CD (SSC, 2004; see Back Cover Insert).

 The analysis presented in this book is based on this substantial body of experience and research. This should provide a sound basis for taking decisions about strategies and policies. However, we are now faced with a problem of imbalance. While we know a lot about the impact of Starter Pack, we know less about the impact of other food security interventions that have been less fully or impartially investigated. Thus, there is a concern that we may favour Starter Pack simply because we know that it works, although other programmes might be prioritized over Starter Pack if we knew more about them. While this concern is a legitimate one, it would be unfortunate if it were to be used as an excuse for inaction. We should ensure that other interventions are also properly researched – but we cannot afford to sit back and do nothing in the meantime. Nor would it be sensible to favour programmes about which there are doubts and which have not been thoroughly investigated over those which have been proven to be effective.

The Contributors

This book is a collection of essays by four broad categories of people: those who invented the concept of Starter Pack and designed the original programme; those who implemented it; members of the M&E teams who researched its impact; and a group of academics and consultants who were invited to critique the programme. In addition to those who contributed directly to it, the book represents a major investment of time by thousands of smallholder farmers in Malawi who were interviewed by the M&E teams. We have done our best to reflect the farmers' views and hopes for Starter Pack. For them, this is not simply an academic exercise. As one village leader pointed out (M. Kakonda, Machinga, Malawi, 2001, personal communication): 'We have been discussing serious things: discussing free inputs means discussing hunger'.

Who is missing from this book? While many individuals made invaluable contributions to Starter Pack and the M&E programme (see Acknowledgements) there is one 'missing voice' that deserves special recognition: that of the critics of Starter Pack, many of whom continue either to misinterpret it or to proceed on ideological grounds instead of looking at the evidence (see Chapter 17). They have played an important role in persuading the contributors that the book should be written. We hope that by making the evidence available to a wider public, it will be possible to have an informed debate on whether or not Starter Pack has a role to play in fighting hunger in developing countries.

How this Book is Organized

This book has four parts. Part 1, 'Origins and Management of Starter Pack', tells the story of how the programme was designed and implemented. It looks at Malawi's problem of chronic under-production of maize because of poor soil fertility and lack of good quality inputs, examines the technical and policy options for dealing with this problem and explains why Starter Pack was conceived as the solution. It considers who and what were the main driving forces behind Starter Pack in terms of policies and politics. It looks at how the programme was managed and how the logistical difficulties of such a large-scale operation were gradually overcome. Finally, it examines the evidence on vouchers and on whether private retailers should be used for pack distribution. This part of the book is intended to provide those who may be considering implementing a Starter Pack programme with a framework for design, pointers on political, institutional and policy issues, and practical guidance on logistics and management.

Part 2, 'Methodology of the Evaluation Programme,' outlines the methods used to produce the evidence presented in Part 3. Given the large scale, complexity and innovative approaches of the M&E programme, there is likely to be considerable interest in this part of the book. It looks at the modular design of the evaluation programme, the building of effective partnerships between local and external consultants and the groundbreaking approach to combining quantitative and participatory research methods. It highlights the problems of fieldwork in Malawi and the lessons learnt about how to achieve consistent, reliable data sets which are comparable across time and space. Finally, it examines the experience of managing a large-scale, multi-module research effort and providing the right financial and other incentives to ensure timely, high-quality research.

Part 3, 'Lessons from Starter Pack', focuses on the policy lessons learnt from the evaluation programme. All but one of the contributors[4] were members of the M&E teams between 1999 and 2003. The policy discussions covered by this part of the book include: How did Starter Pack work, and why was it successful? Why was the impact of Starter Pack so much less when the programme was scaled down? How do farmers decide what to produce and what incentives and constraints do they face? Does the programme crowd out emerging private suppliers of agricultural inputs? Why did community poverty targeting fail? Does Starter Pack undermine sustainable agriculture, or can it contribute to it? How does agricultural extension fit into the picture? Why was the winter season programme much less effective than the main season programme? And how does Starter Pack compare with alternative food security interventions in terms of direct cost (burden on public finances) and indirect cost (adverse macro-economic impact)?

Part 4, 'By Special Invitation', comprises four essays by distinguished academics and consultants, all of whom have a long-standing connection with Malawi but who have had little direct involvement with Starter Pack. They assess the programme from different perspectives. Topics covered include the relationship between poverty, AIDS, food crises and Starter Pack; the broader food security strategy agenda and the role of government and donors; the relationship between free inputs programmes in Malawi, cross-border trade and livelihoods in neighbouring countries; and the question of whether rural development requires a totally different approach based on more interventionist policies, rather than free-market reform accompanied by Starter Pack.

Notes

[1] Self-sufficiency here implies that the country produces enough food for its population to be food secure, but not necessarily self-sufficiency at the household level. Some households may not produce enough themselves, but are able to access food in the market at affordable prices.
[2] There were also small-scale distributions of free inputs in the 2002 and 2003 winter seasons (see Chapter 14).
[3] Safety nets would not normally be considered in this context, but were included in the analysis because some members of the donor community see them as an alternative.
[4] Chris Garforth, author of Chapter 13, was not a member of any of the M&E teams; because of his knowledge and experience of agricultural extension in Africa, he was invited to draw together the findings from several of the M&E studies for the purposes of this book.

References

Mann, C.K. (1998) Higher yields for all smallholders through 'Best Bet' technology: the surest way to restart economic growth in Malawi. *CIMMYT Network Research Results Working Paper* No. 3, Harare, Zimbabwe.(Unpublished, available on CD in Back Cover Insert.)

Munro, L.T. (2003) Zimbabwe's Agricultural Recovery Programme in the 1990s: an evaluation using household survey data. *Food Policy* 28, 437–458.

National Economic Council (2000) *Profile of Poverty in Malawi, 1998: Poverty Analysis of the Malawi Integrated Household Survey, 1997/98*, Lilongwe, Malawi.

SSC (2004) The monitoring and evaluation archive of Malawi's Starter Pack and Targeted Inputs Programmes 1999-2003. Statistical Services Centre, The University of Reading, UK. (Available on CD in Back Cover Insert.)

I

Origins and Management of Starter Pack

1 The Origin and Concept of the Starter Pack

MALCOLM BLACKIE AND CHARLES K. MANN

1.1 Introduction

Since the 1980s, Malawi's food security has rested upon the productivity of fertilized hybrid maize. In the wake of a disastrous drought and a new political climate, the credit programme that had delivered fertilizer and hybrid seed to smallholder farmers was collapsing in 1994. Hybrid maize produces roughly twice as much maize per unit of nitrogen applied as local maize. The government feared disastrous consequences for the food supply should the area planted to fertilized hybrid maize fall sharply. To try to quantify the potential impact of such an event, the government's Food Security and Nutrition Unit developed a simple spreadsheet 'what if' analysis to forecast maize production under a series of assumptions about inputs and weather (Mann, 1994a,b). Under a wide range of plausible assumptions, the model projected serious food shortages.

The model's simple and compelling arithmetic logic persuaded the World Bank and the European Union to support the government's request to forestall a collapse in maize production through a large-scale free distribution of hybrid seed and fertilizer in the 1994/95 main agricultural season. The World Bank's Malawi Deputy Representative, Peter Pohland, summed up the situation for donors: 'We really hate this sort of giveaway input programme, but unless someone has a better idea, we see no alternative but to implement it. Without it we will later face a massive need for emergency food aid' (donor meeting, Lilongwe, July 1994). With the government's Early Early Warning Model continuing to forecast maize production shortfalls (Mann, 1995), free hybrid seed and fertilizer distributions continued the following year. When they stopped, maize production fell precipitously. In 1997/98, Malawi experienced a dire food crisis.

Unlike the situation in 1994, Malawi now had improved technology for smallholder farmers at hand. In 1998, the government's agricultural research

©CAB International 2005. *Starter Packs: a Strategy to Fight Hunger in Developing Countries?* (ed. S. Levy)

and extension team had just completed 5 years of farm-level field testing of an improved maize productivity package tailored to the country's diverse agricultural regions. With data from 1700 farmer field trials (six treatments per trial), the team could identify and recommend 'Best Bet' seed, fertilizer and legume rotations for all smallholder farmers.

Three critical elements formed the core of the recommendations. First, fundamental to improving productivity were two new varieties of semi-flint hybrid maize, developed expressly for Malawi's conditions: MH17 and MH18. These embodied the hard endosperm of the traditional varieties, important to protect maize in on-farm storage from weevils and also to produce good flour from hand-pounding in villages. Also, in contrast to traditional varieties that produced only 8–12 kg of maize per kg of nitrogen, the improved hybrids produced 20–25 kg of maize per kg of nitrogen. The improved output/input ratio (the rate at which plant nutrients are converted to food) was crucial, given that the country's low levels of soil nutrients were severely limiting its maize production capacity.

Second, the recommendations recognized the importance for all farmers (especially the poorest) of maximizing the use of locally available fertility sources. This meant growing leguminous crops in combination with maize and other food crops as well as using farm manures as composts, as these provide organic sources of nitrogen, enhance soil structure and reduce soil erosion. Implementing nitrogen-rich rotations not only reduces the need for expensive commercial fertilizer but also improves diets by adding protein and energy-rich foods such as soybeans and groundnuts.

Third, the 'Best Bet' recommendations reflected economically viable fertilizer doses tailored to regional soil conditions. These adjustments resulted in levels of commercial fertilizer far lower than the previous 'one-size-fits-all' recommendations, and also very different in composition to account for regional soil variation.

The next stage was to move from recommendations to adoption. In the absence of a viable credit programme, how could this new technology be made accessible to all of Malawi's smallholder farmers? While the 'Best Bet' practices were profitable at existing input and output prices, very few smallholders had the cash needed to purchase even small amounts of the required inputs. Thus, the technology for a major productivity breakthrough would remain out of the reach of most of Malawi's smallholders for the foreseeable future.

It was the frustration of this 'so-near-and-yet-so-far' dilemma that gave rise to the concept of giving to all farmers a Starter Pack of the new inputs. It seemed clear that if all smallholders could try the 'Best Bet' package even on a small plot (0.1 ha), national maize production could be increased by around 180,000 t, equivalent to the capacity of the entire silo complex of the Strategic Grain Reserve (Mann, 1998).

As Starter Pack's objective was to bring about long-term change in farming practices, the programme, along with its important complementary measures (an extensive radio extension campaign, packaging commercial fertilizer in small bags, a fertilizer for work programme, efforts to expand savings and credit), was designed to run for 5–10 years. It was not conceived as an emergency one-shot increase in food production. The long-term nature of the programme recognized that it would take time to effect changes in farming systems.

1.2 Poverty, Soil Fertility and Food Security

Even when Malawi was producing a maize surplus (until the mid-1980s), household food security was poor – as indicated by widespread and pervasive malnutrition and one of the highest levels of child mortality in the world. This was largely because agricultural growth was driven by tobacco production in the estate sector and by the introduction of fertilized hybrid maize (using subsidies to disguise the real cost of production) to the wealthier farmers in the smallholder sector. Growth across the agricultural sector was highly uneven and increasingly the result of agricultural export growth in which smallholders played a minor role. Agricultural liberalization policies introduced from the mid-1990s undermined the illusion of price stability, and put an end to the input subsidies and credit programme, which had helped shore up maize production at an increasingly unaffordable cost.

As the fields of Malawi's smallholders became less productive and their landholdings grew smaller, maize became increasingly dominant in the cropping system. This is because in such conditions farm households favour high-calorie crops that can meet their energy needs, such as maize, and reduce the land allocated to other crops. By 1998, maize occupied 85% of smallholder cropland.[1] Rotation crops and intercrops had declined in importance or disappeared. Because of pressure on land, most farmers grew maize in continuous cultivation rather than under the traditional long fallow rotation, which restored soil fertility and reduced the build-up of pests and diseases. As the soil resource base was degraded, crop yields fell.

External influences further damaged Malawi's food security. In the late 1980s and the early 1990s, Malawi's food supply was reduced by drought, and demand was increased by the influx of refugees from the civil war in neighbouring Mozambique. Two imported pests – the cassava mealybug from the Democratic Republic of Congo (then Zaire) in 1973 and the cassava green mite from Uganda – threatened cassava production, an important alternative staple food crop (Herren and Neuenschwander, 1991).

By the second half of the 1990s, Malawi faced an impossible dilemma. The staple food, maize, was largely produced by smallholders, but high input costs locked most smallholders into low-productivity, low-input maize cropping systems that failed to provide enough to feed the household (S. Carr, Zomba, Malawi, 1996, personal communication). These same households could not afford to buy the food necessary to fill their deficits in production, and had to resort to trading their labour for food. Typically, this meant doing *ganyu* on wealthier neighbours' lands during the critical farming periods of planting and weeding – which meant that their own crops were planted and weeded late, further reducing yields.

Food purchases used almost all of the tradable resources (labour and cash) of low-income, food-deficit rural households, leaving little to spare for other needs such as housing, education or farm inputs. With fewer and fewer smallholders able to afford farm inputs, the national harvest fell and surpluses evaporated. Malnutrition, already at unacceptably high levels, was rising further. The end point was, at best, a steady decline in already unsatisfactory nutrition and living standards, and, at worst, widespread starvation.

Neither the Malawi Government nor the donors had foreseen how fundamentally the twin events of the collapse of the credit system (see Chapter 10) and the increased cost of fertilizer (see Introduction) would cripple maize production in Malawi. Once the use of the improved maize seed and fertilizer technology were no longer obtainable on credit and were priced beyond the means of most smallholders, and 'emergency' distributions of free inputs ended, the outcome was tragic. The village-level purchase price of maize quadrupled in 1997/98, producing widespread hardship amongst the poor majority of the population. The liberalization of markets – considered to be essential to Malawi's future growth – was rapidly becoming discredited amongst the public by the high consumer price of maize and by the conspicuous rents being extracted by private traders. The economy was experiencing all the downside effects of liberalization, but few of its benefits.

The deteriorating food security situation in early 1998 threatened to undo the impressive progress made in establishing a policy framework for growth. High maize prices were creating powerful inflationary pressures, compromising household food security, promoting labour unrest and fuelling demands for higher wages. Emergency maize imports contributed to government's runaway expenditure, which further fed inflation. Interest rates rose sharply and the currency collapsed, undercutting productive investment and further driving up the cost of fertilizer for the next crop.

1.3 The Driver of Change: the Maize Productivity Task Force

There had been a consensus among national analysts, donors and international research agencies that the key to solving the food security problem lay in the widespread adoption of improved, resource-efficient, agricultural production technologies – particularly for maize (HIID, 1994b). In 1996, a group of individuals in Malawi – scientists, economists and policy makers – pooled their skills to address the country's severe food shortages. The Maize Productivity Task Force (MPTF) was drawn from both public and private sectors. It liaised with key donor agencies and drew on external expertise and advice. This unique Malawian-led initiative to develop a national consensus on a key policy issue could and should have been sustained. With strong support from the government, it achieved much during its short existence. However, support from the donor community ranged from lukewarm to antagonistic because of the emphasis on maize (see Chapter 17).

1.3.1 The policy options

The MPTF looked at two conventional policy options and found both to be inadequate in Malawi's circumstances:

- *Cash cropping*. Increased income from the sale of cash crops could enable smallholders to purchase the necessary inputs for maize. The successful smallholder burley initiative had been linked to improved maize production on the farms involved, but too few of the poor were able to participate in the growing of burley tobacco. Other cash crops – cotton, chillies, paprika – could serve a similar purpose. But even at generous estimates of involvement in cash crop production, only some 20% of the farming population would benefit.
- *Subsidies*. General price subsidies and credit schemes for inputs were discredited. The beneficiaries of such initiatives were mainly wealthier farmers who were already purchasing the fertilizer and seed. Since most of the fertilizer purchased was used on tobacco, it was estimated that the fertilizer subsidy would benefit mainly the wealthier tobacco growers and leave some 75% of smallholders unaffected. Also, efforts to develop a viable commercial credit system to support cash crop production would be seriously undermined by the introduction of a parallel subsidized credit system for food crops.

1.3.2 The technical options

The MPTF also reviewed the technical options. It concluded that the success of the germplasm-led Green Revolution in Asia, under very different conditions and with more fertile and uniform soils, had biased the research agenda – not just in Malawi but also in Africa more widely – away from crop nutrition studies towards breeding. The MPTF acknowledged that varietal maize improvement would have, at best, a transitory impact on smallholder farming in Malawi *unless farmers also addressed widespread and evident declines in soil fertility*. While improved varieties would produce more output per unit of input, their potential could not be realized in a low fertility environment. Unless this was actively addressed, the productivity of smallholder maize-based farming systems in Malawi would continue to stagnate or decline.

The absence of suitable options to provide crops with adequate nutrients (soil fertility management) is a central problem for many African smallholders. The MPTF noted that the main nutrient constraint in Malawi smallholder maize systems was nitrogen.[2] There are only three sources for supplying nitrogen in arable farming:

- organic sources recycled from within the cropped area or concentrated from a larger area;
- biological nitrogen-fixation; or
- mineral (inorganic) nitrogen fertilizers.

Giller *et al.* (1997), in a comprehensive review of the options for building nitrogen capital in African soils, observe that to restore soil fertility in the absence of mineral fertilizers needs significant allocation of land for the production of high-quality organic materials. In the densely populated rural areas of Malawi, particularly in the southern region, land scarcity prohibits taking land out of production to restore soil fertility.

The MPTF updated the information on *fertilizer use*. It established a comprehensive national fertilizer verification trial, which showed that over the full range of agro-ecologies in Malawi, good returns in terms of grain from fertilizer use were feasible under farmer conditions. A separate analysis developed area-specific recommendations for the economic use of fertilizer based on current fertilizer and maize prices.

1.3.2.1 The fertilizer problem

Fertilizer technology is one of the oldest and best-researched areas of agricultural science. The science of inorganic fertilizer management is well established and in the developed world, farmers have reached a high degree of sophistication in their use of fertilizer. In Africa, however, the application of this basic science is too often faulty. Fertilizer recommendations typically ignore soil and climatic variations found in smallholder farming areas, are incompatible with farmer resources, or are inefficient. The farmer is encouraged to use more fertilizer than is necessary, but often does not receive trace elements that are of vital importance locally and therefore fails to extract the full benefit from its use. Economic analysis of fertilizer policy in Malawi (HIID, 1994a) suggested that improvements in the efficiency of fertilizer use could transform the economics of fertilizer.

Efficient fertilizer use means using only the fertilizer that is necessary. Scientists have worked out generalized fertilizer recommendations that can be used by farmers working on areas of somewhat similar land. However, the crop response to fertilizer is determined not just by the nutrient status of the soil. Rainfall also has a major influence, and African rainfall patterns are highly unpredictable. Finally, the cost of the fertilizer needs to be balanced against what the farmer can expect to receive when the crop is sold. In Africa, rainfall patterns and the economics of fertilizer use are rarely taken into account.[3]

Malawi was an extreme case of poor recommendations. From the 1970s, there was a single maize fertilizer recommendation for the whole country, which took no cognisance of differing soil types, the price of maize and fertilizer or the circumstances of the farmer. The recommendation was for application of the equivalent of 96 kg of nitrogen per ha – an impossible investment for Malawi's smallholders, who are amongst the poorest in Africa.

1.3.2.2 Reducing the need for inorganic fertilizer[4]

Members of the MPTF undertook studies to examine a wide range of technology options that could complement or reduce the need for inorganic fertilizer.

The rotation of maize with grain legumes such as soybean, groundnut or Bambara groundnut was found to be one of the more promising technological options to improve soil fertility for Malawian farmers. Unfortunately, in terms of addressing farm family nutrition, the lower-yielding legumes are the ones with the greatest benefit in terms of soil fertility.[5] Nevertheless, the MPTF's analysis showed that maize–legume rotation using legumes such as soybean and groundnut could have an important role to play in establishing household food security in Malawi, as well as comparing favourably with unfertilized continuous maize in terms of economic net benefit (on a cash basis).

Intercropping, where two or more crops are grown mixed together on the same ground for all or much of their life cycle, is an age-old practice in traditional African food agriculture. However, the soil fertility benefits of most common intercrops are limited. Only small amounts of nitrogen are returned to the soil by intercropped legumes when they are shaded by the associated cereals (Dalal, 1974; Manson *et al.* 1986). Under smallholder management in Malawi, the potential for soil fertility enhancement was known to be very modest. Legume intercrops produce little biomass (and consequently little nitrogen), because they are planted at low populations and suffer from competition from the maize crop and from weeds (Kumwenda *et al.* 1993; Kumwenda, 1995). The MPTF, however, recognized the potential of long-duration pigeonpea, which is widely intercropped with maize in southern Malawi. This is a tall plant that is able to grow successfully without competing with the associated maize crop. It sheds its leaves in the field as it matures, leaving a nitrogen-rich biomass. In addition, pigeonpea grows very slowly and has a deep tap root, which acts as a 'biological plough' to penetrate the hoe pan, which impedes rooting in many hand-hoe cultivation systems.

The MPTF examined a range of other potential legume options such as *green manures*, taking *leaf biomass* from other areas on to cropping land, and the increased use of *composts and manures*. Soil losses under low-productivity cropping on sloping land can be high. *Zero or minimum tillage systems* can reduce soil erosion and the labour required for weeding. The MPTF recommended that research into these areas should continue, but that they were not at a stage that they could be recommended for promotion.

The use of *animal manure* is common in many farming systems to improve soil fertility. However, as pressure for arable land rises in low-productivity African cropping systems, cropping encroaches into areas previously used for grazing. This can bring about a decline in the availability of animal manures as livestock are squeezed out. For example, manure from cattle and other animals can be important for farmers in Zimbabwe, but is rarely available in Malawi. Even in better-endowed Zimbabwe, the supply and quality is inadequate to maintain soil fertility on its own.

The evidence showed that *leaf litter from trees* could make a significant contribution in areas close to woodlands – but deforestation associated with the demand for arable land and for building and firewood worked against this option. *Composted crop residues* were used in wetter areas and where crop biomass production was relatively high. However, composts were rarely sufficient for more than a modest part of the cultivated area and did not provide the productivity boost needed. Crop wastes can be incorporated back into the soil to recycle nutrients. In southern Malawi, smallholder farmers incorporate some residues into the soil, or leave some on the soil surface. In most parts of the central and northern regions, the burning or removal of crop residues is common to release nutrients quickly (although nitrogen and sulphur are lost into the atmosphere), to ease land preparation and to reduce pests and diseases. But the nutrient boost from these practices was found to be insufficient to replace losses through cropping over time.

Thus, the MPTF review showed conclusively that there were few reliable technical options for addressing the soil fertility issue. Malawian smallholder farmers had neither enough cash to buy sufficient fertilizer, nor enough land to supply sufficient high-quality organic materials to boost maize yields. However, by the efficient use of small amounts of inorganic and organic materials combined with a more productive seed variety, they should be able to increase their maize productivity significantly. The MPTF concluded that fertilizer must be an integral part of improving smallholder maize productivity in Malawi.

1.4 Political Support

In early 1998, the highest levels of the Malawi Government recognized that there was a food crisis. There was an active high-level Food Security Committee chaired by the Ministry of Finance (MoF), with the Ministry of Economic Planning and Development (MEP&D) as a key partner. In addition, the Cabinet Committee on the Economy was asking for information on how to increase national maize production. In a serious effort to engage local expertise to find potential solutions, the government asked the MPTF to provide it with a set of recommendations to arrest declining soil fertility and improve the food supply. Without this very powerful signal from government that radical new ideas were being actively sought from amongst its own skilled personnel, what eventually became the Starter Pack concept would never have gained support. Good technical recommendations do not make their way into policy or implementation unless there is cross-sectoral technical engagement and real support by politicians.

The Minister of Agriculture, Aleke Banda, who had earlier been Minister of Finance, recognized that producing sufficient quantities of the national food staple was the key to economic stability and political acceptance. Elias Ngalande, Principal Secretary in the MoF, commissioned the MPTF to develop its Starter Pack proposal into a formal strategy for the government, with donor support, to encourage and empower farmers to implement the 'Best Bet' technologies. The support of Dr Ngalande and Mr Banda was crucial in bringing the Starter Pack proposal into national policy. One night, Mr Banda locked key Ministry of Agriculture (MoA) officials and senior representatives of the major fertilizer companies into the Ministry Conference Room and told them not to come out until the programme was designed properly. Demonstrating support at the highest level of government, President Bakili Muluzi participated in the development of a training video for field staff involved in the programme.

1.5 A Comprehensive Smallholder Strategy[6]

The MPTF concluded that nothing would help quell inflation and dispel the widespread state of gloom and insecurity as much as a bumper maize harvest shared by all of Malawi's farmers, and delivered to consumers in the form of lower maize prices. It decided to seek a way to use the promise of the 'Best Bet' technology to jump-start maize production for all smallholders. This would simultaneously improve the food security of food-deficit rural households and sharply increase the marketed surplus available to urban consumers. The focus was to be on two complementary and widely adoptable options:

- increasing access to the improved maize seed and fertilizer technology; and
- diversifying the cropping system through the adoption of locally suitable combinations with grain legumes, principally as rotations.

1.5.1 The Starter Pack

The original proposal for the Starter Pack involved tiny packs of maize seed and fertilizer: a 2.5 kg bag of hybrid seed – enough to plant 0.1 ha – and bags of the recommended quantity and type of fertilizer,[7] as well as a bag of complementary nitrogen-fixing legume seed. With a conservatively estimated maize yield of 1800 kg/ha, replacing local unfertilized maize yielding 800 kg, the household would gain an extra 100 kg of maize on the 0.1 ha of fertilized hybrid maize. This incremental production would feed a household for more than a month in the hungry season. Its hungry-season market value in 1998 was estimated at a minimum of MK500, which was more cash income than a poor family would see in a year. It was, therefore, a meaningful contribution to family welfare at the household level. In addition, the household would benefit from the legumes produced from the seed included in the pack.

The MPTF proposed that the target group be the entire smallholder population (one pack per household), because from a national point of view, introducing the improved maize seed and fertilizer technology into all zones and to all smallholder farmers should have a high pay-off. Moreover, excluding larger smallholders from the programme would eliminate the participation of many innovative farmers and community leaders. However, the programme excluded more than 30,000 larger farms classified as 'estates'. By using the smallholder definition for beneficiaries, the programme was broadly poverty-targeted.

At the national level, an estimated 1.8 million smallholder households producing 100 kg more maize per household would provide incremental national production of 180,000 t.[8] The MPTF calculated the cost of a representative pack at US$18 per household. With the estimated 1.8 million households, this would make a total direct annual programme cost of US$32.4 million. If the value of the incremental maize exceeded US$180 per tonne, the programme would more than offset its cost. Properly resourced, the programme could be done quickly and on a large enough scale to change the national mood of despair over the deteriorating food situation. The technology was known; the need was urgent.

Unlike a conventional price subsidy, all resources committed to the programme generated incremental production. Also unlike a conventional subsidy, Starter Pack was designed to reinforce the effective operation of the liberalized market. The vast majority of smallholders were so short of cash that, at that time, they represented no market for hybrid seed or fertilizer. Giving them a Starter Pack would not displace commercial purchases. On the contrary, giving them experience with high-quality inputs would stimulate the incentive to purchase more inputs in the long run.

Moreover, incremental food in the hungry season reduces the need to seek *ganyu*, freeing up time for weeding, which is critical to increasing yields. Unlike free food, Starter Pack would reward initiative and good husbandry, especially timeliness of planting, fertilizing and weeding. Yields can be three to four times higher with good husbandry than without. This cumulative process should generate resources to support purchasing small but increasing quantities of hybrid seed and fertilizer.

1.5.2 Complementary measures

The MPTF also proposed a series of other interlinked and complementary elements:

- Ensuring that supplies of small bags of hybrid seed and fertilizer (1–3 kg) were readily available for purchase in all rural markets at prices comparable to those of inputs sold in existing large bags (50 kg).
- Supporting the drive to improve productivity with both traditional extension work and an extensive radio campaign reinforcing the 'Best Bet' extension messages included in the Starter Packs (see Chapter 13).
- Providing opportunities for able-bodied individuals to increase their purchasing power for seed and fertilizer through a structured fertilizer and seed for work programme to be implemented during the dry season.
- Building an effective savings club movement tied to the purchase of agricultural inputs along the lines of that which had proved so successful in Zimbabwe.

1.6 Conclusion

In summary, the Starter Pack programme was a focused programme, intended to be long term, aimed at enabling the poorest to access the improved technologies they needed to break out of the vicious cycle of poverty in which they were trapped. It provided all smallholders with a highly cost-effective means of testing improved maize seed and fertilizer technology with a complementary legume rotation under their own conditions, without the risk inherent in purchasing the necessary inputs. It provided a nationally implemented, but individually operated, technology testing and demonstration programme for a small part of each farm, facilitating farmer experimentation with promising technologies. It provided a unique opportunity to build on the best of local knowledge and expertise (at both farmer and researcher/policy maker level) and represented a practical working example of how to link research, extension and national policy.

In Malawi, Starter Pack arose because of a combination of circumstances:

- The development by the late 1990s of 'Best Bet' technology recommendations for solving Malawi's problem of low and declining maize productivity;
- Widespread and pervasive poverty amongst smallholder farmers that locked them into a poverty trap, unable to access high-productivity inputs;
- The country's shift from maize surplus to severe food shortages, and the urgency felt by policy makers and politicians to create a better way of reaching the rural poor with the improved technologies they needed to break out of poverty;
- A forecasting model, accepted by key donors, that estimated maize production under various weather and input scenarios, demonstrating in advance of planting, potential food crises resulting from sharp declines in fertilized hybrid maize; and
- The existence of the MPTF, which acted as a catalyst for change.

Looking beyond the content of the original Starter Packs and the 'Best Bet' technology, the programme concept had the potential to be developed, refined, and adapted in future years to 'fast track' further technology choices into the smallholder sector. Thus, it could be used to diversify farming systems and increase smallholder incomes.

However, it should be noted that after 2 years of high national maize production in 1998/99 and 1999/2000, the programme concept was changed under donor pressure from its original 'Best Bets' productivity focus to become a targeted safety net package distributing lower-productivity inputs based on open pollinated variety (OPV) maize seed rather than MH17 and MH18 hybrid seed. While subsequently again expanded to all smallholders, the programme continued to favour OPV rather than the content appropriate to extend nationally the 'Best Bets' technology identified by the MPTF. It has yet to be accepted as a long-term national extension programme for high-productivity technology. The MPTF itself no longer meets. The loss of what was a highly effective Malawian initiative to link agricultural research with national policy is tragic – and incomprehensible in an environment where relevance and impact of research are rightly seen as a central objective of technology delivery systems.

Notes

[1] Kumwenda et al. (1996).
[2] This does not discount the importance of other nutrients but recognizes the dominating effect of nitrogen in Malawi (and many other African) smallholder cropping systems.
[3] An illustration of the cost of this neglect is shown by the work of Piha (1993) in Zimbabwe. Piha designed a simple, practical and farmer-friendly system to apply fertilizer based on rainfall patterns and nutrient need. Over a 5-year period, Piha's system gave 25–42% more yield and 21–41% more profit than did the existing fertilizer recommendations.
[4] This subsection draws heavily on Blackie et al. (1998).
[5] One promising exception is the freely nodulating, promiscuous soybean varieties such as the Zambian-developed variety Magoye, which can add some 20 kg of nitrogen per ha, although the availability to the following crop may be compromised by the high lignin content of the residues (S. Mpepereki, Harare, 1998, personal communication).
[6] This section draws heavily on Blackie et al. (1998).
[7] Blackie et al. (1998) suggested that the same amount of fertilizer (10 kg of 23:21:0+4S basal fertilizer and 10 kg of urea top-dressing fertilizer) be distributed in all Starter Packs to facilitate pack assembly. Information was to be provided with the pack on how 'the rate of application should be modified to provide the greatest benefits to the farmer, given his or her location, resources and production aim, as well as the soil texture of the maize field'. This 'representative' Starter Pack provided 6.9 kg of nitrogen for 0.1 ha (the equivalent of 69 kg/ha).
[8] Editor's note: In the event, the incremental production due to Starter Pack was higher than these forecasts predicted (see Chapter 8). The MPTF also underestimated the number of households in rural Malawi (see Chapter 6) and overestimated the cost of a representative Starter Pack. Thus, the total number of beneficiaries and output per beneficiary turned out to be higher than expected, while the total programme cost was lower (see Chapter 15).

References

Blackie, M., Benson, T., Conroy, A., Gilbert, R., Kanyama-Phiri, G., Kumwenda, J., Mann, C., Mughogho, S., Phiri, A. and Waddington, S. (1998) Malawi: soil fertility issues and options – a discussion paper. MPTF/Ministry of Agriculture and Irrigation, Malawi. (Unpublished, available on CD in Back Cover Insert.)

Dalal, R. (1974) Effects of intercropping maize with pigeon pea on grain yield and nutrient uptake. *Experimental Agriculture* 10, 219–224.

Giller, K., Cadish, G., Ehaliotis, C., Adams, E., Sakala, W. and Mafongoya, P. (1997) Building soil capital in Africa. In: Buresh, R., Sanchez, P. and Calhoun, F. (eds) *Replenishing Soil Fertility in Africa*. Special Publication No. 51, Soil Science Society of America, Madison, Wisconsin.

Herren, H. and Neuenschwander, P. (1991) Biological control of cassava pests in Africa. *Annual Review of Entomology* 36, 257–283.

HIID (1994a) Fertilizer policy study: market structure, prices and fertilizer use by smallholder farmers. Harvard Institute for International Development and the Office of the President and Cabinet, Government of Malawi, Lilongwe, Malawi.

HIID (1994b) Trickle-up growth: a development strategy for poverty reduction in Malawi. Harvard Institute for International Development for the Government of Malawi, Lilongwe, Malawi.

Kumwenda, J. (1995) Soybean spatial arrangement in a maize/soybean intercrop in Malawi. Paper presented at the

Second African Crop Science Conference for Eastern and Central Africa, 19–24 February 1995, Blantyre, Malawi.

Kumwenda, J., Kabambe, V. and Sakala, W. (1993) Maize–soybean and maize–bean intercropping experiments. *Maize Agronomy Annual Report for 1992/93.* Chitedze Agricultural Research Station, Lilongwe, Malawi.

Kumwenda, J., Waddington, S., Snapp, S., Jones, R. and Blackie, M. (1996) Soil fertility management in the smallholder maize-based cropping systems of Africa. In: Eicher, C. and Byerlee, D. (eds) *The Emerging Maize Revolution in Africa.* Lynne Reinner, Boulder, Colorado.

Mann, C.K. (1994a) Maize production outlook for 1993/94 in Malawi: a framework to examine consequences of input shortfalls and possible poor weather, with implications for the supply–demand balance. Economic Planning and Development Department, Office of the President and Cabinet, Lilongwe, Malawi.

Mann, C.K. (1994b) Estimates of total maize production, 1994/95 under a range of input scenarios and proposed nation-wide promotional distribution of hybrid seed. Ministry of Economic Planning and Development, Lilongwe, Malawi.

Mann, C.K. (1995) Early early warning: maize production estimates for 1995/1996 under alternative weather scenarios. *Food Security Working Paper,* Ministry of Economic Planning and Development, Lilongwe, Malawi.

Mann, C.K. (1998) Higher yields for all smallholders through 'Best Bet' technology: the surest way to restart economic growth in Malawi, CIMMYT *Network Research Results Working Paper* No. 3, Harare, Zimbabwe. (Unpublished, available on CD in Back Cover Insert.)

Manson, S., Leighner, D. and Vorst, J. (1986) Cassava–cowpea and cassava–peanut intercropping. III. Nutrient concentration and removal. *Agronomy Journal* 78, 441–444.

Piha, M. (1993) Optimizing fertilizer use and practical rainfall capture in a semi-arid environment with variable rainfall. *Experimental Agriculture* 29, 405–415.

2 The Players and the Policy Issues

HARRY POTTER

2.1 Introduction

The Malawi Government's enthusiasm for the Starter Pack concept as a way of addressing food shortages in 1998 was based on an acknowledgement of its technical expertise and cost-effectiveness, compared with the alternative of importing food and supplying it at subsidized prices to the poor. However, there was also a significant political context relating to the 1999 general election. While this context was favourable for uptake of the programme in the first instance, it raised serious concerns among donors. It also implied that commitment to the programme might be tied to electoral cycles.

This chapter examines the political, policy and institutional background to the decision to implement Starter Pack in Malawi in 1998, and looks at how the subsequent evolution of the policy and political context affected the programme. It suggests that the favourable initial conditions for translating the concept into reality were later eroded. Policy changes in the donor community and lack of institutionalization within government meant that the programme was vulnerable when key individuals who had supported implementation of Starter Pack were no longer in post. The Malawi experience is relevant to other countries interested in free inputs distributions, as it indicates that political, policy and institutional factors should not be neglected if a successful outcome is to be achieved.

2.2 Government Support for Starter Pack

The Maize Productivity Task Force (MPTF), established in 1996, examined the technical and policy options for dealing with chronic food shortages; it recommended a set of complementary measures including Starter Pack (see Chapter 1). This initiative fed into the work of a high-level Food Security

Committee active in 1998, which was chaired by the Ministry of Finance (MoF) and included representation from other ministries, donors and non-governmental organizations (NGOs). However, in April–May 1998, the Committee did not have enough influence to have a substantive impact on the details of programme design for the first year of Starter Pack.

By 1998, the Ministry of Agriculture (MoA) had a growing problem of credibility as an agency into which government or donors should invest. Starter Pack offered the Ministry the opportunity to play a lead role in what would undoubtedly become a very high-profile programme throughout Malawi. This could help it to regain a central position within government. The Minister of Agriculture, Aleke Banda, recognized this very quickly. Through his own efforts and his political influence and powers of persuasion, he was able to mobilize support within government and open negotiations with donor representatives. Officials within the Ministry also recognized that Starter Pack might allow them to demonstrate the value of research findings and provide a focus for extension efforts, to counter criticisms of declining effectiveness of agricultural extension services.

2.2.1 Cost of the programme

The cost of carrying out a nationwide Starter Pack scheme was expected to be considerable (see Chapter 1), but potential objections from the MoF were reduced by the availability of stocks of government-owned fertilizer. Fertilizer stocks were held by the Agricultural Development and Marketing Corporation (ADMARC) and by the Smallholder Farmers Fertilizer Revolving Fund Mechanism (SFFRFM). They had been built up by a combination of government purchases and donor gifts. By making this fertilizer available, the government could address two major issues. First, it could demonstrate serious commitment to support a programme designed to assist national food security, without significant additional expenditure out of the annual budget. Second, it would provide a way of indicating agreement with the IMF and the World Bank on liberalization of agricultural markets, as the liberalization agenda included withdrawal of government intervention in fertilizer marketing, to allow the private sector to develop.

Nevertheless, the cost of Starter Pack was beyond the government's own resources. Discussions were necessary to reach agreement with donors on providing financial support. These discussions took place during the second and third quarters of 1998.

2.2.2 The 1999 general election

In mid-1998, the government of Bakili Muluzi was approaching the end of its first 5-year term under the multi-party system, introduced at the end of the Kamuzu Banda era in 1994. The Muluzi government had declared commitment to supporting the strengthening of democratic processes and improved governance. A food crisis could have had serious implications for the political and social environment in the run-up to the general election, due in mid-1999. The Starter Pack programme offered the government the attractive prospect of reducing the risk of food shortages across the country.

At the same time, Starter Pack could potentially provide a political platform that would favour the ruling party. It is therefore not surprising that the MoA was able to obtain support at the highest political level for the scheme. On the other hand, Malawi's development partners, particularly the major donors, were concerned that the programme should not give undue advantage to particular political groups. At a political level, the scope for any nationwide scheme to be used, at best, as a vehicle for promotion of the party in power, and, at worst, as a means of directly influencing voting habits, is considerable.

2.2.3 Food security policy

In 1998, efforts to develop a comprehensive national food security strategy were at an early stage. The Food Security Committee failed to gain political and financial support and lost the opportunity to develop a strategy. The MoA had developed an Agricultural and Livestock Development Strategy and Action Plan (ALDSAP) document, largely with World Bank support, during the period of high hopes immediately following the 1994 election. This document placed considerable emphasis on technical issues concerned with production and rather less emphasis on marketing and institutional issues. Significant funds were available from the World Bank and other donors to take forward the ALDSAP. However, the slow development of management capacity and failure to ensure flow of funds to operational units resulted in little improvement in services to farmers at field level.

With hindsight, it became clear that the ALDSAP took limited account of the complexity of the agricultural sector and the need for a whole range of agencies and policies to work together to support efforts by the MoA. Recognition of this formed the basis of the broader approach adopted in the work to develop the Malawi Agricultural Sector Investment Plan, later Process (MASIP) at the end of the 1990s. MASIP brought together representatives from government, donors and the NGO community to work on a more comprehensive approach to the improvement of the agricultural sector. The novelty of such cooperation in Malawi and the complexity of the range of issues made progress difficult. Agreement was not reached on the framework for future action until 2001, when attention and effort were abruptly switched to dealing with the emerging food crisis. Although Starter Pack had been discussed within MASIP, the slow progress of MASIP meant that the programme was not firmly embedded in Malawian agricultural development strategy.

Although the lack of a national food security strategy meant that the Starter Pack programme was not established within an overall policy framework, the absence of such a framework could be said to have allowed the programme to develop unhindered initially. However, the stand-alone nature of

Starter Pack soon became a problem. The government did not seek to embed it in a policy framework or to locate it within its Medium Term Expenditure Framework (MTEF). This appeared to donors to indicate a lack of commitment to Starter Pack, weakening the government's bargaining position for allocation of donor resources to the programme. However, there is an alternative interpretation: the government's stance may simply reflect a lack of real commitment to general policy frameworks or to the MTEF, these being seen as hurdles (linked to flow of funds) placed by donors for donor leverage.

2.3 Donor Positions

From the outset, the Starter Pack programme provoked heated debate among donors and even between individuals within donor agencies. The debate was complicated by a frequently changing cast of characters. There was also an absence, until recently, of a clear policy framework agreed between different donors and government to provide a context for the discussion. This book is designed to provide evidence and considered views to assist in moving the debate towards an informed – rather than an emotional or ideological – exchange.

It is important to note that the broader policy issues of governance, vulnerability and poverty reduction were only starting to emerge as the basis for agency programmes in 1998. All donor agencies were beginning to change from the traditional intra-sectoral project approach toward shared agendas, including Sector-wide Investment Programmes. None of the donor agencies in Malawi had fully developed policies for food security to guide the discussions on Starter Pack. Discussions were therefore very much coloured by the experiences of individual agency staff members then in post, rather than strong collective or institutional positions.

In mid-1998, the discussions on the Starter Pack programme between donors and government focused principally on the potential for the scheme to provide a cost-effective way for Malawi to support increased maize production. It was anticipated that a Starter Pack scheme implemented in 1998/99 would provide a major contribution to national food security during the 1999/2000 marketing season.[1] There was general agreement among donors that the scheme would be likely to meet its agronomic targets, as the research basis was sound (see Chapter 1). There was also agreement, even from those like the United States Agency for International Development (USAID) who were against the overall idea, that if the scheme were to go ahead, it should as far as possible be targeted to those most in need. It should not be directed toward those able to access inputs, e.g. members of tobacco-producer groups. Thus, a registration process to identify appropriate beneficiaries would be necessary.

There were disagreements from the outset on issues such as beneficiary dependency, impact on private-sector agricultural-input suppliers and cost-effectiveness compared with other interventions. However, there was a shared concern among donors that if Starter Pack went ahead it should not become a political issue. This meant ensuring accountability, transparency and avoidance of political partisanship in beneficiary selection and distribution of packs. The mechanisms designed to achieve this are outlined in Chapter 3. The evaluations of the programme indicate an acceptable level of success, given the general level of development in Malawi. For instance, the Monitoring Component of the 2000/01 Targeted Inputs Programme (TIP) (Lawson *et al.* 2001) observed that village heads were invited to area meetings where they were addressed by:

> a wide range of district-level figures. By having politicians, government and other key figures all speak with one voice, potential conflict and dissent was kept to a minimum. Particularly important was the involvement of politicians from across the political spectrum, which effectively de-politicised what was potentially a very political issue – especially in the run up to the local elections.

In 1998, all donors felt that beneficiary dependency was a potential danger if free inputs were to be provided year after year. Thus, while government clearly felt that the scheme should be established from the outset as a multi-year programme, the donors, led by the UK Department for International Development (DFID), saw it only as a short-term measure. Its longer-term evolution or replacement would then depend on the development of national strategies for food security and safety net provision, which would be likely to include a range of interventions.

In 1998, the debate on the impact of the private sector was particularly strong between those then in support of the scheme – DFID, the World Bank and the European Commission (EC) – and USAID, the principal donor supporting private-sector development. USAID feared that the programme would undermine efforts to develop private-sector agricultural input suppliers. Supporters considered that those likely to be beneficiaries of Starter Pack would not have the resources to purchase inputs in the absence of such a scheme.

2.4 The Shift in Focus

2.4.1 Safety nets

In the late 1990s, both government and donors began to focus attention on the chronic nature of poverty and vulnerability in Malawi. All parties began to work on a comprehensive strategy to address these issues, with inputs from the World Bank. The Malawi National Safety Net Strategy (NSNS), adopted by government and donors in 2000 (National Economic Council, 2000), saw agricultural input supply as one of four major safety net interventions to support the poorest in Malawi; the other three were Public Works Programmes (PWPs), feeding programmes for vulnerable groups (mothers and young children) and Direct Welfare Transfers (DWTs). The NSNS argued that a mixture of free, subsidized, credit-supported and free market options would be required to improve access to agricultural inputs. Mechanisms for this mixture are still under development and testing at the time of writing (2004), and the balance between them is not yet clear.

In 1998, prior to the broader debate and policy development on safety nets, the options were very limited. PWPs and targeted feeding programmes were restricted in scope and coverage, and DWTs were almost non-existent. For DFID, the World Bank and the EC, the Starter Pack programme, although with acknowledged question marks, was virtually 'the only game in town'. However, by the 2000/01 agricultural season, the policy environment had changed significantly with the finalization of the NSNS. Starter Pack evolved into the Targeted Inputs Programme (TIP), becoming more a safety net than an agricultural production programme. The institutional locus within government also changed. Although the MoA remained a key government stakeholder, the Department (later Ministry) of Economic Planning and Development became more involved as the agency responsible for the safety net agenda.

The NSNS aimed for a balance between interventions in terms of number of beneficiaries, and suggested that free agricultural inputs would be provided to approximately 350,000 households. While donors and government[2] broadly accepted this figure, they recognized the potentially serious negative response from the rural population if the number of packs of free inputs was reduced from universal provision (enough for 2.8 million households) to 350,000 in one go. Agreement was therefore reached to reduce the number of beneficiaries in steps, as the scope of the other interventions was expanded. The numbers adopted for 2000/01 and 2001/02 were 1.5 million and 1 million, respectively. The selection of beneficiaries was to be based on a targeting process involving the community, using vulnerability criteria derived from the NSNS work (see Chapter 11).

2.4.2 Food crisis and scaling up

In 2002/03, the programme was scaled up to a main, donor-funded distribution of nearly 2 million packs ('List A') and an additional, government-sponsored distribution of some 800,000 packs ('List B'). The size of the 'Extended TIP' programme was influenced by the food crisis of early 2002. Donors agreed to a temporary increase in beneficiary numbers. They and the government agreed that this would be a cost-effective way of adding to national food supply through the 2003 harvest, as part of the medium-term plan for recovery from the 2002 crisis. The number of packs provided in 2003/04 was reduced to 1.7 million, given the improved food situation in 2003.

The scaling up suggested that the donors had recognized a proven, cost-effective role for Starter Pack in rapidly restoring national food security after the crisis. However, donors did not accept the argument that a universal or near-universal programme should be established on a multi-year basis. This reluctance was not based on considered appraisal of the evidence gathered through the comprehensive evaluations, but reflected headquarters-led ideologies of the agencies, within which there was little consideration of the special characteristics of Malawi.

By 2003/04, DFID was the only donor funding the programme (see Chapter 17). Decisions about scale continued to be taken on an *ad hoc* annual basis through negotiations between the government and DFID (see Section 2.5.1). The NSNS continued to envisage scaling down to a small TIP, reflecting a fundamental change from the original concept of Starter Pack as a means of promoting new agricultural technology.

2.5 Programme Design Issues

2.5.1 The principal actors and agencies

The annual negotiations on free inputs programmes from 1998 to 2004 involved a very limited number of individuals. The two most consistent principals were the Minister of Agriculture (Aleke Banda from 1998 to 2000 and from 2002 to 2003), representing the Malawi Government; and Harry Potter, the DFID Rural Livelihoods Adviser, representing the main donor. Alex Namaona of the MoA's Planning Division also played a key role in drafting the government documentation and acting as main Starter Pack/TIP Coordinator within the MoA, with the support of successive Principal Secretaries.

Mr Banda and Dr Potter negotiated the size and scope of the programme each year, based on available resources and the prevailing policy environment, including donor agency views. They were largely responsible for ensuring participation of other government agencies and negotiating financial commitments by DFID and other donors. Neither Mr Banda nor Dr Potter is now in post. Without their continued involvement, the programme is likely to enter an uncertain era. As mentioned above, there has been little sign that it has become fully embedded in the emerging policy agenda. Its future is therefore precarious.

In these circumstances, it is likely that political influences may be as much driving forces as any technical (agricultural or safety net) issues. Following the 2004 general election, the nature of the inputs programme was altered by the new government of Bingu wa Mutharika as part of the process of distancing itself from the programmes of the previous regime (see Introduction). The lead role within government shifted from the MoA to the MoF[3].

2.5.2 Programme management

On the programme management side, the key player for the last 6 years has been Charles Clark, Head of the Starter Pack/TIP Logistics Unit. Without his expertise, the practicalities of programme design and implementation would have been impossible.

The Logistics Unit has been key to the implementation of the programme. It has arranged registration of beneficiaries, public awareness meetings, tendering for inputs and services, production of vouchers and packs, delivery and accounting (see Chapter 3). Donors and government agreed on the establishment of this semi-autonomous unit at an early stage. It was felt that this would ensure accountability and efficiency of operation, and there is no doubt that these have been achieved successfully. The Unit has been able to overcome constraints in staffing and skills within the MoA to handle a complex, country-wide operation. The partnership between Mr Clark and Mr Namaona has been particularly productive, an indication of the importance of getting the right people into place to ensure success.

Under Logistics Unit management, procurement of seed and fertilizer has used a flexible combination of donor procurement agencies and government procurement systems. There has been rapid, open tendering without political or other interference, to the satisfaction of both government and donors. As the programme schedule is linked to the onset of the annual rains, speed has been an essential requirement.

If discussions were to lead to a greater embedding of the programme into mainstream government activities, the potential transition of the Logistics Unit into the government structure might become an issue. Fear of lack of transparency in procurement and accounting in a unit entirely within government, based on unfortunate past experience, might affect the level of donor support. For government, the handling of a staff complement with a range of specialized skills only required for part of the year would be likely to create financial and administrative problems; such issues are easier to handle in a separate unit.

The use of a semi-autonomous Logistics Unit accords with the current agenda of redefinition of the role of government to concentrate on policy, regulation, support, coordination and monitoring, rather than implementation. This is a matter to which other countries considering Starter Pack programmes would need to give careful consideration.

2.5.3 Beneficiary numbers and programme coverage[4]

Since 1998, the debate on beneficiary numbers has been the most contentious issue in the design of the annual Starter Pack/TIP. Government was attracted from the start to the idea of 'universal' coverage, based on one pack per rural household. It considered that this would be administratively easy and politically attractive in the run-up to the 1999 general election. The government accepted that this would mean that some households that could afford to buy inputs from their own resources would receive packs. However, it argued that this would still help national food security by adding to maize production.

At the outset, the supporting donors examined the possibility of targeting packs to the poorest areas (geographical targeting). However, it was clear that population and poverty distribution in Malawi would lead to a preponderance of packs going to the southern region, the stronghold of the ruling party. This was felt to be a potential source of political strife, capable of undermining efforts to support a peaceful, fair, multi-party election in 1999. Donors therefore reluctantly agreed that universal coverage in all parts of the country was the only practical solution for the 1998/99 season.

In 1998, even with agreement to provide one pack per rural household throughout Malawi, there were two problems. First, there was disagreement about the number of households in rural Malawi; and second, there was no register of households that would allow managers of the programme to work out how many packs were needed in each location.

Government accepted the need for a registration process to provide a mechanism for an equitable 'one pack – one household' distribution. The registration process was handled by the Logistics Unit, which registered beneficiaries using first the MoA's extension workers and then, from the 2000/01 season, the staff of the District Assemblies (see Chapter 3). But what was the right number of households to register? Preliminary figures from the 1998 census indicated a rural population of approximately 8.5 million, with an average household size of 4.5. This suggested a total of some 1.9 million rural households. However, in 1998, when the Logistics Unit processed the list of names of household heads registered by the extension workers, it found a total of 2.86 million. It was clear that the definition of a household was imprecise, but it seemed that whatever the definition used, in many cases more than one member of a household was being registered. However, the government was reluctant to reduce the figures from those initially registered. After considerable discussion, donors agreed to use a figure of 2.86 million, on the basis that – even with a potential over-supply – national food security would benefit.

The 1999 general election process influenced numbers for the second Starter Pack programme in 1999/2000. Discussions on numbers were initiated prior to the election, and both sides agreed that any change from universal coverage could endanger a peaceful election. However, with the election scheduled for late June and registration carried out in May, local politicians quickly recognized the significance of the registration process and put pressure on extension workers to inflate the registers; when the Logistics Unit counted the number of households, they found 3.7 million. This figure was recognized as unrealistic by all concerned and the registers were pared down to 3.3 million. At this point, financial constraints prevailed, as the donors refused to support a programme of more than 2.86 million packs.

An evaluation study was commissioned in 1999/2000 to estimate the number of households in rural Malawi (see Chapter 6). It estimated 2.78 million households and found fair agreement between the Logistics Unit's register (2.86 million version) and reality on the ground. The Logistics Unit's register, known as the Farm Family Database, has been maintained and updated annually and is now in use by a variety of agencies as a powerful planning tool. The 2004 update includes Global Positioning System (GPS) data, which allows integration with other digitized planning data sets. It was, for example, used to provide a secondary check for voter registration numbers in the 2004 general election.

2.5.4 Pack composition

From the start, the packs contained inputs for 0.1 ha. The contents included 2 kg of improved maize seed, 2 kg of legume seed (later reduced to 1 kg), fertilizer for use at planting (23:21:0+4S basal fertilizer) and urea top dressing fertilizer. Maize was included as the main staple grain in Malawi, with some attempt being made to match the hybrid seed supplied to local growing conditions. The legumes were included for the potential triple benefit of improved protein levels in the household diet; improvement of soil conditions through nitrogen fixation and penetration of compacted soil from hand cultivation by the tap-roots; and a contribution to household incomes as cash crops for which there was robust market demand.

The area-specific seed and fertilizer recommendations of the MPTF could, in theory, have been used to develop area-specific packs. An early attempt to use region-specific cereal seed (see Chapter 12) proved difficult to manage, but there was greater success with the legume component. In practice, it was felt that the logistics of preparing a large number of different packs and ensuring their correct delivery were beyond local capacity and resources, particularly with respect to fertilizer levels. A blanket recommendation of fertilizer application was therefore adopted from the start, based on an assessment of a compromise among the area-specific levels. For the hybrid maize seed, 10 kg of basal and 5 kg of urea were provided.[5]

The shift in emphasis from an agricultural production programme to a safety net intervention in 2000/01 resulted in a change from hybrid to certified OPV maize seed, which can be saved by the farmer for further use without the need for annual free distribution or purchase of seed. Field trial evidence indicated that the yield difference under field conditions between the hybrids and OPVs would be very small, but both would significantly out-yield traditional varieties. The change was intended to ensure that farmers obtained the production potential of improved seeds on a sustainable basis, while benefiting from the fertilizer response. The quantity of basal fertilizer was reduced to 5 kg, as there was some evidence that this would not seriously jeopardize yield levels, while providing a significant saving on input costs for the programme. Full evaluation of the impact of this change has not yet been made, and debate on whether hybrid or OPV maize seed is appropriate for Starter Pack programmes in Malawi and elsewhere continues (see Chapters 1 and 12).

2.5.5 Extension

Every year since 1998/99, the packs have included leaflets with extension messages about how to use the inputs provided. The MoA has also been given resources to provide field extension support, principally through its frontline extension workers, but also through rural radio broadcasting. Both of these components are crucial to support the implementation of the free inputs programmes, but have proven difficult to implement successfully. Starter Pack/TIP is not the only intervention supporting maize production in Malawi and involving the MoA, and each initiative produces its own extension materials. There have been occasions where messages from the various programmes have been in direct conflict on matters such as plant spacing or whether to intercrop or plant pure stands. As a result, farmers (and some extension workers) have been confused. However, senior officials have been reluctant to raise such issues for fear of potential loss of the benefits that they derive from donor support to each of the different interventions. This reflects the limited management capacity within the Ministry and the lack of technical self-confidence of its staff.

The leaflets provided in the packs have been developed to try to maximize the impact of the extension message (see Chapter 13). The evaluations have been used to inform their design, but it is clear that there is still room for improvement. The within-pack messages have been supported by radio broadcasts prepared by the Agricultural Communication Branch (ACB) of the MoA. Technical support from UK-based communications consultants has helped to improve the content and impact of the messages and develop the broadcasting skills of the ACB's radio unit. A welcome spin-off from this support has been the improved capacity of the unit to disseminate information effectively in support of other rural programmes, including land policy reform and forestry management.

Training programmes for agricultural extension workers have been in place for each of the annual Starter Pack/TIPs, but the evaluation programme has indicated a poor level of effectiveness in extension work with farmers (see Chapter 13). Field staff have seen the free inputs programme as yet another demand on their time, with little in the way of specific incentives to encourage their efforts. Impact has also been reduced because of loss of extension workers by attrition, especially from HIV/AIDS, leaving many posts unfilled.

2.6 Conclusion

Efforts were made from the beginning of Malawi's Starter Pack programme to encourage debate with a range of actors in the donor community. However, the debate has been largely dominated by ideological and emotional rather than evidence-based comments. Government has not provided long-term support to the programme or its evolution by its failure to embed the programme into a policy framework or the MTEF. Most donors have therefore retreated from involvement in the programme, or even discussion of it. As the principal actors (including the author of this chapter) are now no longer in post, its future in Malawi is very uncertain.

However, the evidence now available from the uniquely comprehensive and rigorous evaluations (see Parts 2 and 3) provides support for a role for a substantial free inputs programme in Malawi's national food security and livelihood improvement planning. There is no doubt that the Malawi experience has produced a cost-effective, efficient mechanism for smallholders to increase food production, particularly of the staple food crop – maize. The programme also remains a vehicle for demonstration and promotion of improved technologies for crop production and diversification, as the content of the packs and the associated messages can be altered easily to suit specific needs. Whether it should be based on universal or near-universal coverage will depend on its place in Malawi's food security and agricultural strategy agendas, but its proven cost-effectiveness should not be ignored. Issues such as coverage can be settled objectively only when comparisons of options are based on evidence from independent, external evaluation rather than ideology or emotion. The methodologies developed within the Starter Pack programme for gathering a comprehensive range and quality of data over a number of years – with comparability of results from year to year – set a standard for an evidence base. The investment of approximately 1% of total programme costs for evaluation has provided a rich return.

Other countries with problems of chronic food insecurity may benefit from Starter Pack programmes. Any future programmes in Malawi or elsewhere should consider carefully how to tackle the potential problems of political pressure and electoral cycles. They should also attempt to place the intervention within a broader, longer-term policy context. Finally, they should seek to institutionalize it within government and the donor community, while keeping programme management separate under a semi-autonomous Logistics Unit.

Notes

[1] The use of the term 'marketing season' in Malawi refers to the same period as the main agricultural season (from one maize harvest in March–June to the next), but indicates that we are discussing marketing rather than production. For example, maize produced in the 1998/99 main agricultural season is harvested in 1999 and sold in the 1999/2000 marketing season. The food security impact of the 1998/99 Starter Pack would be felt in the 1999/2000 marketing season.
[2] Many MoA officials were opposed to scaling down the programme, but had to concede defeat given that the size of the programme was dictated by funding constraints.
[3] This is intended to ensure closer integration of the free inputs programme into the budget as part of the measures to encourage resumption of direct budget support by donors, following its withdrawal at the end of 2001 due to fiscal management issues.
[4] This section discusses beneficiary numbers for 'universal' Starter Pack only. The numbers for the TIPs (1.5 million, 1 million and the goal of 350,000) are simply percentages of this total.
[5] Owing to resource constraints, the 10 kg of urea recommended in the 'representative' Best Bet technology pack (see Chapter 12) was never provided.

References

Lawson, M., Cullen, A., Sibale, B., Ligomeka, S. and Lwanda, F. (2001) 2000/01 TIP: findings of the monitoring component. (Unpublished, available on CD in Back Cover Insert.)

National Economic Council (2000) *Safety Net Strategy for Malawi*. Lilongwe, Malawi.

3 The Logistics and Costs of Implementation

CHARLES CLARK

3.1 Introduction

This chapter describes how Starter Pack is implemented. First, it looks at the system as a whole and how it is managed. Second, it examines the logistics of distributing packs to large numbers of beneficiaries throughout Malawi. Third, it discusses how much the operation costs to run and presents a cost breakdown. The specific information provided is based on the most recent incarnation of the programme: the 2003/04 main season Extended Targeted Inputs Programme (ETIP). This was undoubtedly the most successful of the six main season distributions that have been implemented since 1998/99. However, the mechanism for implementation has evolved over 6 years, over a steep learning curve. The chapter ends with some reflections on the pitfalls encountered and the lessons learnt during implementation of the programme.

3.2 The System

The main objective of the team charged with implementing Starter Pack is to deliver packs of seed and fertilizer to beneficiaries in all of the 30,000 or more villages located in Malawi's 28 districts. This must be done over a period of roughly 2 months (mid-October to mid-December), so that the inputs are available in time for planting, before the main rains. Beneficiaries should not have to walk more than 5 km to collect a pack. To achieve this, the implementation team must ensure that the right number of packs are delivered to each of 2000 distribution centres throughout the country on the right day.

The system involves eight key areas of activity:

1. Agreement between government and donors on number of beneficiaries and content of packs to be distributed.

2. Selection and registration of beneficiaries.
3. Distribution of one voucher to each beneficiary, entitling him/her to collect a pack.
4. Procuring the agricultural inputs and delivering them to seven warehouses.
5. Packing the bags of seed and fertilizer, together with leaflets providing instructions on the use of the inputs, into 'mother bags' (packs) at the seven warehouses.
6. Transporting the packs to the 2000 distribution centres.
7. Distributing the packs to the registered beneficiaries in exchange for vouchers.
8. Reconciliation of accounts, payments and liaison with funding agencies.

The second part of this chapter describes these activities in detail. First, we look at the management of the system and the players involved in making the system work.

3.2.1 The management

At the centre of the system is the Logistics Unit. One of the myths about Starter Pack is that it requires a bloated bureaucracy. In fact, the Logistics Unit has four professional staff (the Team Leader, two Senior Logistics Officers and a secretary/data entry person) and three support staff (two drivers and a cleaner). Other staff are hired on temporary contracts – warehouse logistics clerks, data entry and voucher counting personnel – while all the work of procurement, packing, transport and delivery of the inputs is contracted out to private sector suppliers of goods and services (see Section 3.2.2).

In 2003/04, the Logistics Unit's IT equipment consisted of a server, three networked computers, a stand-alone laptop computer and three printers. The Unit also had a photocopying machine, a telephone, a fax, an open pick-up truck and a vintage Landrover. All had been supplied by the UK Department for International Development (DFID) in previous years from equipment returned from other projects. The result was that a programme serving 1.7 million beneficiaries throughout Malawi was managed at a cost of 1% of total expenditure (see Section 3.3.7).

3.2.2 The key players

The Logistics Unit coordinates several categories of player:

- District Assemblies (DAs), which handle the selection and registration of beneficiaries and distribution of vouchers.
- Procurement of seed and fertilizer and delivery to the warehouses is subcontracted to private traders through a system of bids supervised by the procurement agent, Charles Kendall & Partners (CKP).

- The Agricultural Communication Branch (ACB) of the Ministry of Agriculture (MoA) Extension Department is responsible for the design of the leaflets providing instructions on use of inputs and their delivery to the warehouses.
- Packing of 'mother bags' at the warehouses is carried out by organizations that have won bids for warehousing, supply of bags and assembling the packs. After a competitive tendering process, seven warehouses are selected: two in the north, two in the centre and three in the south.
- Transporting the packs to the distribution centres and distribution of packs to beneficiaries are carried out by organizations that have won bids for these tasks. Bid documents contain information indicating the numbers of packs to be distributed and the number of distribution centres per Traditional Authority (TA). The bid document for transporters includes indicative mileages.

In addition to ensuring that the players carry out their tasks as required, the Logistics Unit is responsible for reconciliation of accounts and liaison with donor agencies to ensure speedy payment to all involved. It also reports to officials at the MoA.

3.2.3 Key management tools

In order to manage such a complex system with so many players – in particular, to ensure that each player fulfils its obligations before payment is released – it is vital to have good information. The Logistics Unit has developed two key information management tools: the farm family database and a commodity tracking system (CTS). The farm family database stores information on beneficiaries in over 30,000 villages, and is the only up-to-date register of villages in the country (see Box 3.1). The CTS database is capable of generating reports on suppliers' performance, packing progress and uplifts to distribution centres (see Section 3.3.5).

3.3 The Logistics

Figure 3.1 shows the steps required to implement Starter Pack. Commencement is dependent on a Memorandum of Understanding defining the number of beneficiaries to receive a pack (see Chapter 2). This document is prepared by the MoA's Starter Pack Coordinator and sent to prospective donors with the blessings of the Minister. Discussions between donors and the Ministry result in an agreed version stipulating numbers of beneficiaries and pack composition. In 2003/04, the number of beneficiaries was fixed at 1.7 million and the pack was to contain 5 kg of basal fertilizer, 5 kg of urea, 2 kg of open pollinated variety maize seed (OPV) and 1 kg of an appropriate legume. Legumes distributed were soybeans, beans, pigeonpeas and groundnuts. In the case of groundnuts, 1.5 kg were provided to compensate for the fact that shelled groundnuts were distributed.

> **Box 3.1.** The farm family database.
>
> The farm family database template has been in existence since 1999, with each year seeing improvements to its performance. At the beginning of 2003/04, it was capable of absorbing entries covering beneficiary names, villages, wards, TAs and districts. It also contained a field covering the distribution centres used when distributing the packs. In the 2003/04 season, data were entered for 1.7 million beneficiaries in over 30,000 villages.
>
> The database output allows reports to be generated at a number of levels. For instance, a district report lists the names of all TAs in the district and gives the number of wards, villages and beneficiaries in each TA. A village report gives the names of all beneficiaries in the village. Each village is linked to its distribution centre, allowing beneficiary registers to be compiled by the distribution centre and to show the names of all villages and beneficiaries associated with each centre. At the end of 2003/04, a list of all villages and distribution centres was prepared for each district. This was forwarded to the DAs with a request to update the village lists and review the location of the distribution centres in preparation for the following year's distribution.

3.3.1 Selection and registration of beneficiaries

On receipt of the approved national distribution figure, the Logistics Unit advises all 28 districts of the number of beneficiaries that they have been allocated. In 2003/04, this meant dividing the 1.7 million national figure roughly in proportion to rural population because, as in previous years, the government had decided that every village should receive packs. With fewer packs than the total number of farm families (probably some 2.2 m), this meant that just under 80% of households would receive a pack in each village. There was a slightly higher weighting for the southern region, where the proportion of the population seen as needing a pack is greater (see Chapter 11). This was offset by slightly lower weightings for the central and northern regions.

DAs are given blank registers and their staff are asked to write down the names of beneficiaries. Funds are provided to facilitate mobility. Registration should be carried out by task forces containing local representation including traditional leaders, church authorities, government officers and non-governmental organizations (NGOs). Task forces are asked to give priority to selecting the poor as beneficiaries (see Chapter 11).

Once beneficiary registers have been completed, they are returned to the Logistics Unit. A data entry centre is opened, with ten networked computers served by ten data entry clerks. In 2003/04, DFID provided the computers, which were reclaimed from other projects. Data entry commenced on 25 August and finished at the end of October, when updating of the farm family database (see Box. 3.1) was complete. The entry for each beneficiary name is accompanied by the district vehicle registration letters (e.g. BL for Blantyre) and a unique number, running in sequential order. Each name has a unique combination of letters and numbers. Beneficiary registers, organized by distribution centres, are then printed.

Logistics and Costs 45

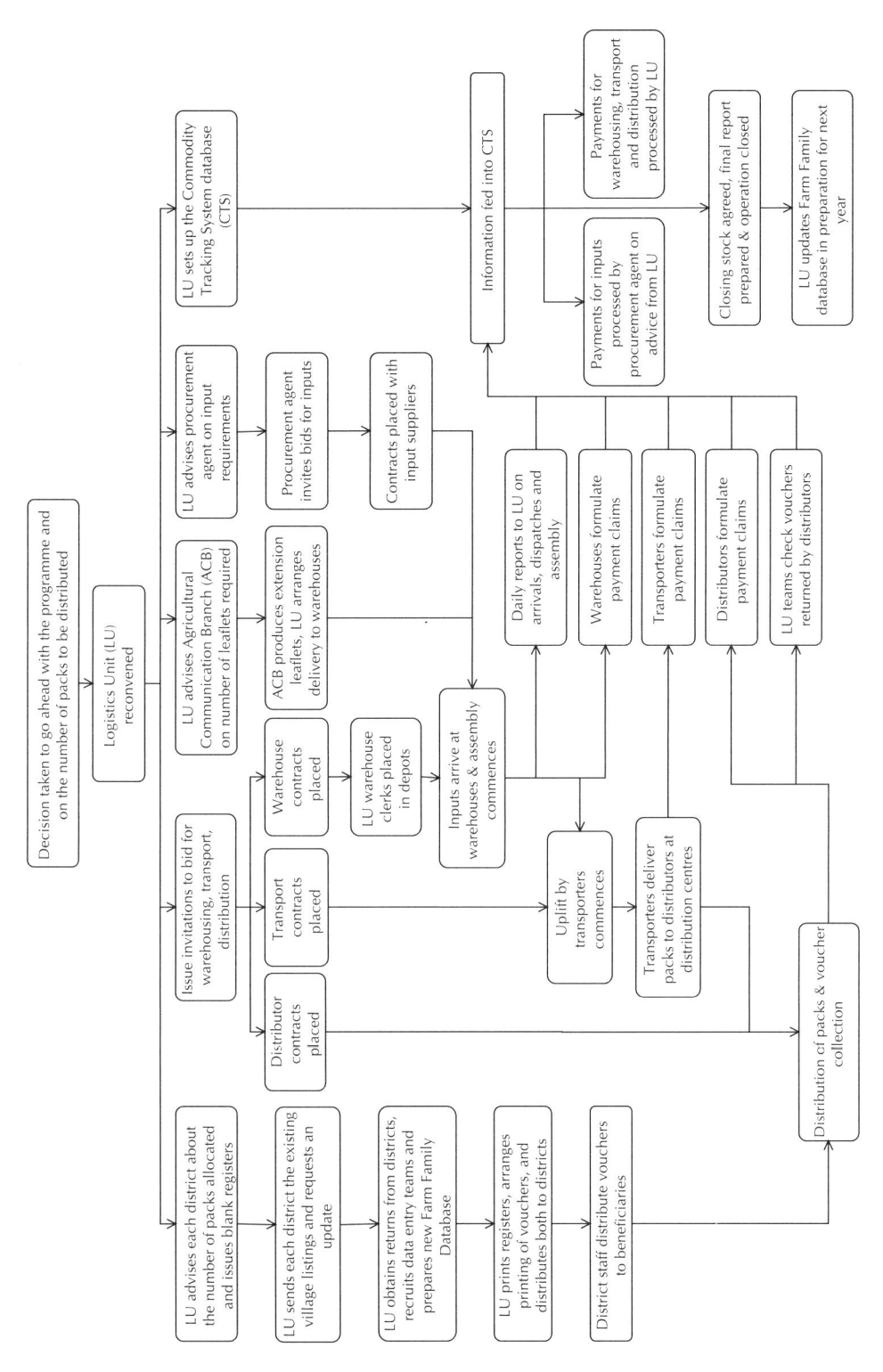

Fig. 3.1. Flow chart of Starter Pack implementation.

3.3.2 Voucher distribution

Meanwhile, local printers are commissioned to print vouchers (1.7 million in 2003/04). The vouchers are packed separately for each distribution centre, with voucher numbers corresponding to those in the beneficiary register for that centre.

A team from the Logistics Unit accompanied by the MoA Starter Pack Coordinator travels to each district for pre-arranged meetings with district officials: the Chairman of the DA, the District Commissioner (DC), the District Agricultural Development Officer (DADO) and the TAs. Vouchers and beneficiary registers are handed out, and the team stresses the importance of every beneficiary receiving the voucher with the same number as the number beside his/her name in the register. An extra copy of each beneficiary register is provided so that a copy can be left in each village; registers should be displayed in a prominent place so that villagers can check the names of the registered beneficiaries. The TAs are responsible for ensuring that all the vouchers and registers reach the villages.

3.3.3 Procurement of inputs

Once the number of beneficiaries has been agreed between the MoA and donors, and the allocation of beneficiary numbers per district has been finalized, the Logistics Unit notifies the procurement agent (CKP) of the quantities of maize, fertilizer and legumes that will be required at each of the seven warehouses identified for the programme.

CKP then issues invitations to bid to traders. As well as quantities, the bid documentation contains reference to packing demands: each commodity supplied must be packed in bags of the right size (e.g. in 2003/04, maize seed had to be packed in 2 kg bags and fertilizer in 5 kg bags). It also stipulates the documentation that a supplier is required to produce as evidence of proper quality control (e.g. certificates of seed quality from the MoA's Research Station at Chitedze). After the bids have been analysed by CKP, the Logistics Unit, the MoA and the donors, suppliers are advised of the extent of their awards.

3.3.4 The leaflets

The ACB is also informed of beneficiary numbers per district. It is responsible for designing and printing the technical leaflets (instructions on how to use the seed and fertilizer) that need be included in the packs. The leaflets are printed in Chichewa for the central and southern regions and in Tumbuka for the northern region. The ACB should deliver the leaflets to the warehouses where the seed and fertilizer are being packed, but in 2003/04 the Logistics Unit arranged for delivery to the warehouses.

3.3.5 Distributing the packs

Once the vouchers are in the hands of the beneficiaries, and the inputs and leaflets have been delivered to the warehouses, the scene is set to commence delivery of packs to the beneficiaries. Two further preparatory measures are needed:

- The Logistics Unit advises the DAs of the transporters and distributors who have won bids to operate in their districts, and all three parties meet with Logistics Unit staff to agree on the distribution schedule for each TA.
- The Logistics Unit places logistics clerks in the seven warehouses and equips them with fax machines and the necessary stationery to control arrivals and dispatches.

3.3.5.1 Control of warehouse arrivals and dispatches

As each supplier truck arrives carrying fertilizer, maize or legumes, the logistics clerk records its status on a Daily Receipts Report, detailing the waybill number, the truck number, the commodity and the weight carried. These reports are faxed each evening to the Logistics Unit, and the information is entered in the CTS database.

The commodities arrive pre-packed in the right size bags. As they arrive, the pallets are broken open and the individual bags are placed in a mother bag. The logistics clerk records the number of mother bags available and the number packed. This information is forwarded to the Logistics Unit on a daily basis, where it is entered in the CTS and used to process interim payments to warehouse contractors.

Dispatches are controlled through the issuing of Authority to Collect forms (ATCs). The logistics clerks have an ATC book, which is printed using carbonized paper. Each ATC form consists of five copies and is divided into three sections.

The *first section* of the ATC form indicates dispatch and is filled in by the logistics clerk. The information to be completed includes the truck number, the amount to be loaded, the TA and distribution centre to be served. A fresh ATC is used for each distribution centre to be serviced by a truck. The logistics clerk leaves the last copy in the book and gives the remaining four to the truck driver, who proceeds to the loading bay, where he gives all four copies to the loading superintendent. When the truck is loaded, both truck driver and loading superintendent sign the *second section* of all four copies of the form, where the loading superintendent has filled in the box indicating the quantity loaded. The loading superintendent then gives three forms to the driver, retaining one.

The driver then proceeds to the distribution centre, where offloading takes place in the presence of the distributor. The distributor signs the *third section* of all three copies of the form presented by the driver, recording the number of bags received. The distributor retains two copies and returns the third copy to the driver. The driver's copy of the completed ATC becomes a proof of delivery (POD), which the transporter presents to the Logistics Unit with his invoice and tabulated mileage as support for payment.

Meanwhile, back at the warehouse, the logistics clerk copies all the information from the first section of each ATC onto a Daily Dispatch Report and faxes it to the Logistics Unit for entry in the CTS database. The CTS now has enough information to generate reports covering suppliers' performance, warehouse stocks and deliveries to distribution centres. This allows the Logistics Unit staff to check and process invoices from suppliers, warehousemen and transporters when these are presented. Payment is made by the donors on recommendation from the Logistics Unit. In the case of commodity suppliers, CKP is required to ensure that each claim for payment is accompanied by the necessary quality control documentation (see Section 3.3.3).

3.3.5.2 Control of distribution
This section describes how distribution should take place if instructions are followed, and the controls that have been put in place. However, it should be noted that the Monitoring and Evaluation (M&E) teams have detected a number of irregularities which result in some registered beneficiaries failing to receive their packs (see Chapter 4). The Logistics Unit is still trying to close loopholes which can be exploited by the players, but does not currently have sufficient monitoring/policing capacity to carry out this role (see Section 3.4).

Distribution of packs to beneficiaries normally takes place immediately on arrival of a truck at a distribution centre. Distributors are instructed that they should advise village leaders 72 h in advance of distribution, so that beneficiaries can assemble at the distribution centre. On arrival, beneficiaries should be grouped by village. The distributor should call out the name of each beneficiary from the register, check the number of his/her voucher against the corresponding name and number on the register and release a pack in exchange for the beneficiary's voucher if (and only if) the numbers agree.

The distributor should then group all vouchers by distribution centre in sequential number order and present the vouchers together with a copy of the POD and invoice to the Logistics Unit. The vouchers should be counted carefully by the Logistics Unit, missing numbers recorded and any losses deducted. A recommendation for payment should then be made to the donors. In 2003/04, distributors were paid between MK19 and MK22 per voucher collected, and there was a penalty of MK450 for any packs that could not be accounted for (i.e. the distributor failed to submit a voucher and did not return the pack to the warehouse).

3.3.6 Finalizing

With the payments to the distributor, the main tasks in the programme are complete. It only remains to agree on any residual stock in the warehouses, arrange for its transfer to safe custody and to write up the final report. The final report from the Logistics Unit for the 2003/04 programme (Logistics Unit, 2004) includes the following statistics:

- 1,699,110 beneficiaries were registered.
- 1,682,853 packs were distributed.
- 1,680,725 vouchers were recovered.
- 2128 packs were distributed to individuals who had lost their vouchers. These distributions were authorized by the DAs.
- 6864 packs could not be accounted for and penalties were charged to transporters or distributors at the rate of MK450 per pack.

3.3.7 Cost of the programme

Table 3.1 presents a breakdown of the costs for the 2003/04 ETIP. It shows that 81% of programme expenditure went to beneficiaries in the form of inputs (fertilizer and seed). Logistics Unit management costs comprised a mere 1% of total expenditure. With 1.7 million beneficiaries, the cost of each pack (or cost per beneficiary) was US$7.

Of the total cost of MK1286 million (US$11.8 million[1]) in 2003/04, the government contributed MK108 million in the form of fertilizer, and the Logistics Unit raised MK650,000 from the sale of tender documents. All other funds came from DFID.

Table 3.1. Breakdown of costs for ETIP 2003/04.

Commodity/activity	MK	% of total
Fertilizer	626,495,274	48.72
Maize seed	282,968,501	22.00
Legume seed	134,248,457	10.44
Registration	10,855,730	0.84
Voucher costs	10,717,674	0.83
Pack distribution	36,389,274	2.83
Transport	38,001,392	2.96
Warehousing and bagging	63,369,707	4.93
Logistics Unit costs	14,229,799	1.11
CKP agency fees	21,398,581	1.66
MoA (administration)	355,810	0.03
MoA (leaflet production)	3,714,144	0.29
General project support	33,524,719	2.61
HIV/AIDS support	6,077,780	0.47
Related food security issues	3,607,726	0.28
Total	1,285,954,568	100.00

Source: Logistics Unit/DFID accounts.

3.4 Lessons Learnt

To reach the present level of efficiency, many changes have been made and a number of hard decisions have been taken. This section highlights some of the lessons learnt.

3.4.1 Beneficiary registration

In the first 2 years of Starter Pack, agricultural extension workers (field assistants) were responsible for selecting beneficiaries. This role made them unpopular and undermined their capacity to act as providers of extension support to farmers. Since 2000/01, the DAs have played a much greater role and the creation of task forces charged with responsibility for beneficiary registration has reduced the pressure on agricultural extension workers. Local government structures have proved capable of handling registration, but there is still much resentment on the part of those excluded from the registers. The transparency of the process should be increased, and both beneficiaries and non-beneficiaries should have a means of checking the registers, which should be displayed publicly in every village.

3.4.2 Management

The original idea was that the MoA would manage Starter Pack. However, it soon became clear that Ministry staff had too much work to do. Moreover, government procurement of the inputs led to delays and controversy about the tendering process. It is now clear that a dedicated, autonomous unit like the Logistics Unit – with the support of both government and donors – is necessary to manage such a complex programme. However, the presence of an enthusiastic Starter Pack Coordinator in the MoA, who forms a link with policy makers, is also vital.

3.4.3 Timing

A key factor for a successful operation is timing. In the 2000/01 main season, problems with procurement of inputs (still in the hands of government at this stage) led to late delivery of the packs to farmers, severely eroding the benefits of the programme.

A decision to go ahead early in the year is essential to ensure timely delivery. A lead time of 5 months is required between deciding numbers of beneficiaries and commencing distribution of packs. In 2003/04, approval in principle was given in May, although the final decision on numbers was not taken until July. This enabled tendering processes to be put in place, allowing contracts for inputs to be awarded in July. Consequently, distribution could commence in October and was concluded in November before much rain had fallen.

However, to be certain of the right inputs being available (particularly OPV maize and legume seed), a clear signal that there will be a project should be given 12 months before distribution. Only then will growers feel confident enough to grow the required seed.

3.4.4 Sub-contracting and payments

Sub-contracting has contributed to a 'lean and mean' operation. However, it took 3 years for all parties to accept that competitive tendering for input procurement, packing, internal transport and distribution produced the most efficient and cost-effective result.

There has also been much discussion about who should coordinate transport between warehouses and distribution centres, with 'transport brokers' hired to handle this in the first 4 years. However, this system tended to break down at crucial moments. From the 2002/03 season, management of transporters has been in the hands of the Logistics Unit.

To keep suppliers of goods and services on board, speedy payment facilities have had to be arranged. The programme has also had to make special financial arrangements for the impecunious DAs, who need cash advances to carry out their work.

3.4.5 Logistics

The following lessons have been learnt about logistics:

- There should be a watertight bidding system and contract documentation to cover supply of commodities, warehouses, transportation and distribution.
- Availability and quality of agricultural inputs should be fully researched before the placing of supply contracts.
- Inputs should be pre-packed by suppliers in bags of the right size.
- Adequately sized warehouses with good loading bay facilities are needed.
- Realistic targets for packing and loading should be developed for each warehouse.
- Trained logistics clerks, reporting to the programme managers, should be placed in every warehouse.
- There is a need for a transport pool with sufficient capacity and versatility of truck size to move the packs to the distribution centres.
- There should be well-designed systems for issuing ATCs to transporters picking up packs from warehouses to take to distribution centres.
- Truckers delivering packs to distribution centres should not be provided with a second load without producing proof of the first delivery.
- Retention should be held on payments to suppliers, warehousemen, transporters and distributors until tasks are completed and the final payment can be safely certified.

3.4.6 Distribution of packs

The following problems have arisen with pack distribution. First, if penalties for 'lost' packs are not strictly applied, there is an incentive for distributors to sell some packs, as their market value is higher than the MK19–22 which the distributor receives (but lower than the MK450 penalty charge). Lack of penalties for transporters provides similar incentives. Penalties were applied strictly in the 2003/04 ETIP after this problem was identified, and this should send the right signals in the future. Penalties must be adjusted as input prices rise, so that they remain higher than the market value of the packs.

Second, the system of numbered vouchers which match numbers on beneficiary registers (distribution control system) depends on three main elements:

- The integrity of those carrying out the registration and voucher distribution, so as to avoid corrupt officials registering themselves and giving themselves vouchers with which to collect packs. At present, the registration and voucher distribution process is entrusted to local government (district) officials.
- All beneficiaries turning up at the right time with the right vouchers, and distributors being there to deliver the packs. In reality, errors and delays may mean that the distributor ends up just collecting the right total number of vouchers for the packs distributed rather than matching voucher numbers with the names and numbers on the beneficiary register. Distributors face no financial penalty for discrepancies between voucher and register numbers, and such penalties would not be practical as it is difficult to apportion blame in such cases.
- The Logistics Unit's capacity to count the vouchers returned by distributors, to record missing numbers and investigate any problems that arise.

It is natural that a distribution system on a scale as large as Starter Pack presents such challenges. However, the current system has almost no in-built monitoring or policing functions and limited capacity to investigate problems. A modest amount of independent monitoring/policing would undoubtedly improve the performance of the system.

3.4.7 Information systems

It is clear that such a large-scale operation as Starter Pack requires databases that can cope with registration of beneficiaries, movement of commodities and distribution of packs. This may present challenges if such databases do not exist prior to the establishment of the programme, but the Malawi case shows that they can be successfully developed. Once a database has been established, resources should be allocated to keeping it up to date.

The Logistics Unit also discovered that a newsletter is a useful way of keeping players informed and, more importantly, aware of what actions are required of them. The management team produced a weekly newsletter with a wide circulation list covering donors, politicians, input suppliers, service providers and civil servants.

Notes

[1] Using the end-2003 exchange rate of MK108.566:US$1.

References

Logistics Unit (2004) Final Report: implementation of Targeted Inputs Programme 2003, Lilongwe, Malawi. (Unpublished, available on CD in Back Cover Insert.)

4 Pack Distribution and the Role of Vouchers

Anthony Cullen and Max Lawson

4.1 Introduction

In the first 2 years of Starter Pack, the programme was universal – every smallholder household was to receive a pack. The system for distributing packs to beneficiaries was simple: beneficiary registers (lists of names) were compiled for every village, the required number of packs for the villages in the surrounding area were calculated for each distribution centre and this number of packs was handed out on the appointed day, by calling out the beneficiaries' names from the register and asking them to claim their packs.

With the introduction of the Targeted Inputs Programme (TIP) in the 2000/01 main season, a large number of former beneficiaries were to be excluded from the programme (see Chapter 2). As there were only enough packs for 1.5 million smallholder households,[1] two concerns arose:

1. Would the 'commotion' (disturbances at distribution centres) reported in previous years worsen if whole villages turned up to collect packs, but there were not enough packs to go around?
2. How could the programme managers ensure that the 'right' households (those registered as beneficiaries) received the packs?

To deal with these problems, the use of vouchers (printed pieces of paper entitling the bearer to collect a pack) was proposed, and the system described in detail in Chapter 3 emerged. First, beneficiary lists were compiled by local government officials working through village task forces. Then the Logistics Unit printed a voucher for each beneficiary named on the list, and these vouchers were handed out before the day of pack distribution. On the day, those with vouchers could go to the distribution centre to collect a pack – for which they had to hand over a voucher. Finally, distributors could then claim payment from the Logistics Unit by submitting the vouchers collected from beneficiaries.

©CAB International 2005. *Starter Packs: a Strategy to Fight Hunger in Developing Countries?* (ed. S. Levy)

Theoretically, this meant that the problems of commotion and of misallocation of packs should be solved because:

1. Nobody without a voucher would turn up at a distribution centre, so there would be no fighting over packs; and
2. Assuming that the registered beneficiaries all received a voucher, the packs would go to the right households.

However, the problem of misallocation of packs continued. The 2001/02 TIP evaluation survey detected some evidence of it, so rigorous quantitative investigations were organized in the 2002/03 main season and the 2003 winter season to check on 'mismatches' between registered beneficiaries and recipients of packs (see Box 4.1). They found that a large percentage of registered beneficiaries were failing to receive packs and around 10% of packs were 'leaking' from the system before even reaching the villages (see Section 4.2.1).

In 2003, in response to the evaluation findings, the Logistics Unit introduced a distribution control system. This is based on giving each voucher a unique serial number that matches a name and number on the beneficiary register. The Logistics Unit also tightened up payment of financial penalties by distributors for missing vouchers. These changes appear to have helped somewhat, but there are still a number of loopholes in the system (see Chapter 3).

This chapter reassesses the arguments in favour of and against vouchers as a method for controlling distribution of benefits in a large-scale programme like Starter Pack.[2]

In the first part of the chapter (Section 4.2), we present the evidence of problems. This evidence comes from the Monitoring and Evaluation (M&E) programme and from the authors' interviews with key stakeholders following the 2003/04 main season Extended Targeted Inputs Programme (ETIP). We consider whether the existing system can be tightened up, and whether this is worthwhile or whether there are better alternatives. Our focus is on ensuring that the registered beneficiaries receive the packs, as this should be the main objective of the voucher system.

In the second part of the chapter (Section 4.3), we consider another possible reason for using vouchers. It has been argued that vouchers would allow private retailers to take over the distribution of Starter Pack, thus stimulating the private sector. Although not an official aim of the programme, this private sector development argument is widely felt and hence merits analysis. The advantages and feasibility of channelling pack distribution through retailers is examined in some depth. However, we consider this to be a secondary objective of Starter Pack: food security and the interests of the beneficiaries should be paramount.

4.2 Do Vouchers Help in Distributing Packs?

4.2.1 The evidence

In 2000/01, two M&E studies investigated whether vouchers helped to reduce 'commotion' at distribution centres. Chinsinga *et al.* (2001) found that 'People endorsed the voucher system, because without the vouchers, the distribution

Box 4.1. How the data on 'mismatches' was collected.

The survey teams carried out field verification to establish whether the packs were received by those on the beneficiary register. A careful procedure was followed to check this:

- Villages were selected using a multi-staged sampling scheme: 138 villages for the 2002/03 main season TIP evaluation and 90 villages for the 2003 Winter TIP evaluation (Nyirongo et al., 2003; Levy et al., 2004).
- In each village, the survey team compiled a list of all households by going from door to door collecting information about each household. This included the name of the household head and any other person who had received a TIP pack, and whether the pack was received in that village or in a different village. The survey team then compared the list of beneficiaries found in the village (on the household listing form) with the beneficiary register (the 'Targeted Inputs Programme Logistics Unit (TIPLU) list'). Anyone found in both was marked as 'Found in first round of visits and TIPLU list'.
- Frequently, some of the names on the TIPLU list were not found. The survey team then went through a process of checking these names. This started by sitting with the village leaders and reading out – one by one – the names of people on the TIPLU list who had not been found. If the village leaders said they recognized the name, an attempt was made to match it to one of the names in the household listing form. If a probable match was found, the team then went back to the household to verify if that individual really existed and lived in the village. If the match was confirmed, the individual was classified as 'Verified in the second round'.
- In some cases, individuals named in the TIPLU list could not be found. Such cases were marked as 'Not found'. In other cases, individuals appeared on the household listing form as having received a pack although they were not on the TIPLU list. These cases were marked as 'Not on TIPLU list but received packs'.

With the information from the field verification exercise, a statistical analysis was carried out to establish the percentage of packs correctly allocated to registered beneficiaries and the percentages of different types of misallocation. The large number of villages visited allowed us to estimate the number of 'mismatches' with a reasonable level of confidence.

Source: Carlos Barahona.

exercises would have been quite disorderly'. Lawson et al. (2001) found that 'Vouchers were widely felt to have a very beneficial impact on security, as only those with vouchers tended to come to the distribution centre on the most part'.

However, the evidence on allocation of packs to registered beneficiaries was less favourable:

- The evaluation of the 2001/02 TIP found that 19% of the packs distributed did not reach the intended recipients (Levy and Barahona, 2002).
- The evaluation of the 2002/03 TIP found that 25% of the packs distributed did not reach the intended recipients.[3]
- The 2003 Winter TIP survey found that 18% of the packs distributed did not reach the intended recipients and 10% went missing before even reaching the villages. It also found a serious problem of 'ghost' names: 13% of names on the beneficiary register could not be found on the ground (Levy et al., 2004).

Two issues are important here. First, packs being reallocated within the village so that they end up with non-registered households instead of registered households. This is a serious matter for a programme based on poverty targeting, as poorer households may be losing out to wealthier ones. However, in a universal (untargeted) Starter Pack programme, within-village reallocation should not give too much cause for concern. On the other hand, the second issue – packs failing to reach the villages at all – is a serious problem for any programme, as it suggests 'capture' by people who are not part of the broad target group (smallholder farmers).

If vouchers do not ensure that packs go to registered beneficiaries, then part of the rationale for having them is undermined. It could even be argued that vouchers help to create more opportunities for abuse of the system. Levy and Barahona (2002) argue that:

> . . . as most of the leakage is associated with the voucher distribution, the problem was probably caused in the main by the village and/or district authorities, who were responsible for distributing the vouchers. It would not be difficult to give some of the vouchers to people whose names did not appear on the registers, particularly in view of the secrecy surrounding the beneficiary selection process.

Vouchers have a clear monetary value: they can be swapped for packs containing valuable seed and fertilizer, and they are needed by distributors to ensure that they receive payment. Therefore, it is not surprising that those charged with distributing vouchers may wish to keep some for themselves. One way that officials can do this is to create ghost names on the beneficiary register, keep the ghost names' vouchers and then approach distributors at or after distribution to exchange the ghost names' vouchers for packs. Chinsinga *et al.* (2004) found many ghost names in the 2003 Winter TIP villages, some of which had apparently been used for personal advantage by Ministry of Agriculture (MoA) employees who were providing technical assistance. Levy *et al.* (2004) found that the survey data provided support for 'the hypothesis that some of the ghost names may have been created by corrupt officials for the purpose of keeping some of the Winter TIP packs for themselves'.

4.2.2 Can vouchers work?

Despite the problems described in the previous section, it would be premature to do away with vouchers. They clearly have advantages for security and when reinforced by financial penalties, they are a useful management tool for verifying delivery of packs by distributors and controlling payments to distributors (see Chapter 3). But the problem of failure to ensure delivery to the right beneficiaries remains. The Logistics Unit has made steps towards solving this problem, by the introduction of the control system based on giving each voucher a unique serial number that matches a name and number on the beneficiary register. However, there are still some challenges on the side of beneficiary registration and voucher distribution – tasks carried out by the District Assemblies (DAs).

The Logistics Unit has repeatedly recommended the appointment of an independent body to distribute vouchers (Logistics Unit, 2001, 2002, 2003). However, this would undermine the role of the DAs, which is crucial to the success of beneficiary registration and to maintaining the farm family database (see Chapter 3).

A compromise might be to introduce a high-profile system of voucher distribution monitoring by carrying out extensive spot checks on whether the vouchers have been distributed correctly before pack distribution commences. In addition, serious efforts should be made to reinforce the Logistics Unit's capacity to investigate problems with vouchers; and there should be a well-publicized system of penalties for local government and MoA officials and others found appropriating vouchers that do not belong to them.

An independent voucher distribution body would go a long way to addressing the problems, as would a monitoring system with effective penalties for abusing the system. Either or both of these measures should be utilized to close the loopholes that currently provide opportunities for packs to be diverted to persons not registered as beneficiaries.

Irrespective of which approach is employed, the process of beneficiary selection should be made more transparent, with lists of registered beneficiaries displayed publicly in every village to promote accountability (Levy and Barahona, 2002).

4.2.3 Comparing the alternatives

As the voucher system for pack distribution has a number of drawbacks as well as advantages, we need to consider whether the continued use of vouchers is the most appropriate means of implementing a large-scale pack distribution. Would another method be better? Table 4.1 compares four alternatives. The current approach of the Logistics Unit is half-way between Alternatives No. 3 and 4: a number-based control system has been introduced, but it is not yet fully supported by a system of spot checks and penalties.

The answer to the question 'Which method is best?' needs to take into account at least two key factors. First, the context in which the programme is to be implemented. If the context is one of extreme poverty and weak government capacity, as in Malawi, the distribution of valuable commodities like seed and fertilizer will be subject to pressures and abuse. This may be compounded by a lack of transparency and accountability. In such contexts, leakages are to be expected without strict controls. The second factor to take into account is what level of leakage is permissible. In some cases, the managers may consider that some leakage of the benefit is acceptable – in which case Alternative No. 2 or 3 is probably the most desirable. In other cases, as with Starter Pack in Malawi, the managers will want to minimize leakages – in which case they should consider Alternative No. 4. However,

Table 4.1. Comparing alternative approaches to distributing packs in large-scale programmes.

	Alternative No. 1	Alternative No. 2	Alternative No. 3	Alternative No. 4
Basic characteristics	• Packs distributed without beneficiary registration or vouchers	• Registration of beneficiaries • Packs distributed to those on register, without vouchers	• Registration of beneficiaries • Vouchers distributed to those on register • Packs distributed to voucher holders	• Registration of beneficiaries • Vouchers distributed to those on register • Packs distributed to voucher holders • Use of number-based control system backed up by spot checks on voucher distribution and financial penalties for distributors
Purpose/benefits	✓ Useful for relief and emergency operations	✓ Useful for programmes where beneficiaries need to be identified	✓ Specific beneficiaries are identified ✓ Distributors held to account with vouchers ✓ Civil disturbance minimized	✓ Same as Alternative No. 3, but also helps to minimize abuse of voucher system
Risks/problems	✓ Packs particularly vulnerable to theft and misallocation ✓ Potential for civil disturbance	✓ Complex and time-consuming process ✓ Potential for civil disturbance ✓ Process open to some abuse, particularly invention of 'ghost' beneficiaries	✓ Complex and time-consuming process ✓ Voucher system opens up more possibilities for abuse – creation of 'ghosts'	✓ Heavy demands on time and resources because control system accompanied by spot checks and penalties is used to check abuses

the cost of a number-based control system supported by spot checks and penalties should be carefully considered, as it may outweigh the benefits in some cases.[4]

4.3 Do Vouchers Open the Door for Retailers?

A number of retailers argue that the distribution of packs through retail outlets would be cheaper than the current Starter Pack distribution system, and would also provide an important stimulus to the private sector and thus to Malawi's economic growth.

Vouchers are seen as providing the opportunity for retailers to take over distribution from the existing players. The proposed system envisages continuing to register beneficiaries and distribute vouchers as described in Chapter 3. However, the remaining stages of the process – procurement of inputs, warehousing, transportation and distribution – would be handled by retailers, who would 'sell' packs of seed and fertilizer to beneficiaries in exchange for their vouchers. The retailers would submit the vouchers to the Logistics Unit for payment, as distributors do under the current system.

The second part of this chapter evaluates this proposal, both in terms of the programme's primary objective of ensuring that beneficiaries receive their packs of seed and fertilizer (at the lowest possible cost), and also in terms of the secondary objective of stimulating the private sector.

4.3.1 Can retailers distribute the packs?

Involvement in Starter Pack would provide a considerable source of extra revenue to retailers. Thus, it is not surprising that they strongly support the idea. One of the arguments advanced in its favour is that distribution would cost less if managed by retailers. As retailers already have staff employed at their outlets, the cost of recruiting for the distribution exercise would be saved. However, this presupposes that distribution would take place solely from the location of existing retail outlets. One retailer envisages the process as follows:

> Instead of taking it to the school we would take it to one of our shops . . . Instead of dropping it off in five different places we would drop it off in one and the people would come to that place on a certain nominated day. They would present their voucher and walk off with their mother bag. As a retailer, we would also charge for this service.

Although retailers service every district in Malawi, they are thin on the ground in many areas (see Fig. 4.1). In places with few outlets, beneficiaries would have to walk distances in excess of 30 km to receive a pack. Taking into account the hardship that would be caused by travelling such distances,[5] and the probability that it would lead to reduced uptake of the packs by beneficiaries, the option of conducting distribution exclusively through retail outlets should be rejected. According to a senior MoA official, retail centres tend to be:

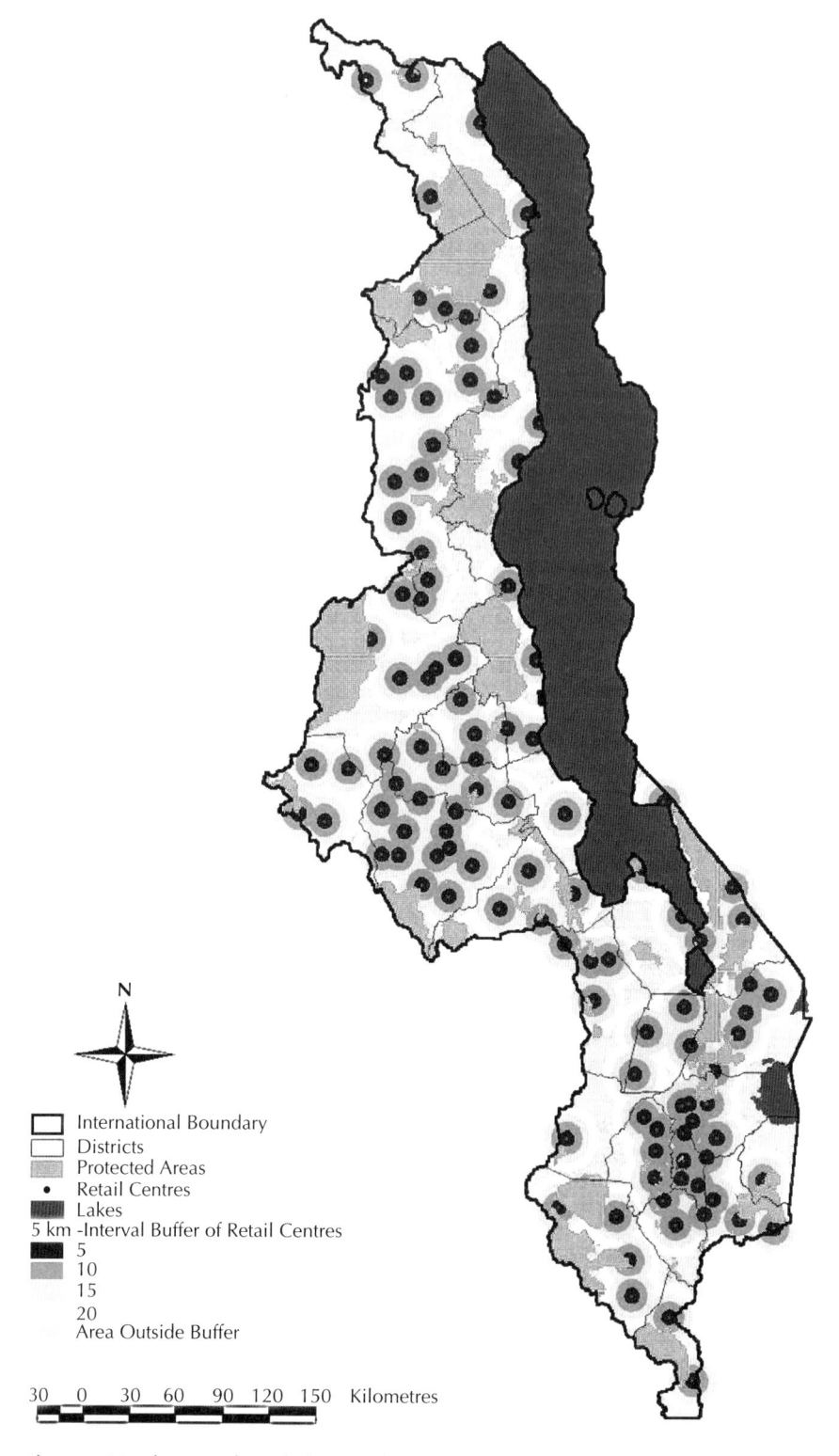

Fig. 4.1. Distribution of retail chain outlets in Malawi. (Rook and Maleta, 2001.)

... concentrated along the main roads and rarely set up in rural areas. The infrastructure does not exist to support retailers in these areas. It does not make sense in terms of profitability to place outlets in such places ... The potential for the use of vouchers in the supply of packs through retailers is extremely limited and would have to be confined to selected areas.

Even if the question of reasonable access for the beneficiary is put aside, it is unlikely that the use of retailers would be cheaper than the current system. It was estimated by one of the larger retailers interviewed by the authors that the cost of transport would make up an average of 55–60% of the cost of the final product. In addition, there would be the expense of inspections by the Malawi Bureau of Standards (1.5–2%) and mark-ups charged by the retailers for each pack (5–10%). The final cost would also have to take into account the serious exchange rate risk of importing fertilizer. As fertilizer is imported on credit and paid for in foreign currency, the retailer is vulnerable to losses if the Malawi Kwacha depreciates and he cannot increase the price of the packs to compensate – as would be the case if he signed a contract with the Logistics Unit to provide packs at a fixed price. The combination of transport costs, inspection costs, mark-up and a margin to cover exchange rate risk would clearly have a significant impact on the cost of the final product.

In addition to the retailers' costs, the programme would also have to budget for monitoring the service provided, as retailers are prone to try to squeeze extra profits out of such opportunities. In a recent pilot study of direct welfare transfers in Dedza district, Malawi, retailers agreed to sell goods to project beneficiaries in exchange for vouchers; M&E teams discovered that the retailers often raised the prices of their goods, because the beneficiaries had no option other than to spend their vouchers at the designated retail outlets, which were few and far between (Levy *et al.*, 2002). In the case of a programme where vouchers are exchanged for packs of agricultural inputs, price increases would not be possible – but it would be possible to short-change beneficiaries by providing less fertilizer and seed in each pack, or to charge them a small cash fee in addition to the vouchers.

An idea of the cost of involving retailers can be obtained from a comparison with the IFDC Sustaining Productive Livelihoods through Inputs for Assets (SPLIFA) project, which in 2003/04 provided packs of 10 kg of maize seed and 50 kg of fertilizer (urea only) to 30,000 beneficiaries through 70 retailers – ranging from large retail outlets to small dealerships. The retailers did not import or transport the inputs; they only received them, stored them and distributed them to beneficiaries in exchange for vouchers, which were then collected for payment by IFDC staff. For this service, the retailers charged around 1 US dollar ($) per voucher.[6]

By contrast with a retailer-based system, the Starter Pack system of delivery managed by the Logistics Unit lowers costs by using economies of scale and competitive tendering. The cost of purchasing inputs is reduced by buying in bulk and competitive tendering. The cost of warehousing and transportation and distribution of packs is reduced through the competitive bidding process. This exploits competition in the private sector and helps lower the cost of pack delivery. In the past, retailers have taken part in the competitive bidding for distribution contracts, but none of the major retailers have won contracts.

We conclude that in Malawi it is unlikely that by using retailers it would be possible to meet the primary objective of getting packs to all eligible beneficiaries, or that if such a system were used it would be cheaper than the current one. However, this would not necessarily be the case in other countries. Where the rural population has better access to retail outlets and transport costs and exchange rate risks are lower, the involvement of retailers in pack distribution may well be feasible. Close monitoring would be required in the early stages, to pick up any abuses and design ways of preventing them.

4.3.2 Would the involvement of retailers stimulate the private sector?

The private sector is already involved in all stages of the process of pack delivery managed by the Logistics Unit: input suppliers, warehouse providers, transporters and distributors (see Chapter 3). This arrangement provides a significant stimulus to the private sector. The analysis of the previous section suggests that transferring contracts to retailers would provide a boost to large retail chains at the expense of other private agents, which run more competitive businesses. We would argue, therefore, that any proposal to distribute packs using retailers would have to demonstrate that they were a competitive alternative to the service provided by the businesses that are involved at present.

4.4 Conclusion

Vouchers are a worthwhile option to consider as a tool for distributing packs of free inputs on a large scale. However, the original idea of the 2000/01 TIP that vouchers alone would ensure that the benefits went to the right households (registered beneficiaries) proved naïve. In fact, vouchers create additional opportunities for abuse of the system in the form of capture of the benefit by people outside the target group. If they are to be used in a Starter Pack programme, they need to be reinforced by a number-based control system backed up by spot checks on voucher distribution and financial penalties for distributors. Beneficiary selection processes should also be made as transparent as possible.

An alternative to a control system with monitoring/policing and financial penalties is to appoint an independent voucher distribution body. However, this would threaten the relationship of the programme managers with local government structures (DAs).

One argument in favour of vouchers is that they provide an opportunity for retailers to take over pack distribution. Retailers argue that this would stimulate the private sector and economic growth. However, the evidence shows that Malawi's retailers are not in a position to take over pack distribution at present: their geographical coverage is poor in rural areas and the cost of distributing packs through retailers would be high. Moreover, the private sector is already involved in all stages of the delivery process: the Logistics Unit contracts out most of the key functions to private players (input suppliers, warehouse providers, transporters and distributors). These private sector agents are more competitive than the country's large retail chains. Nevertheless, in other countries the involvement of retailers may be feasible and desirable. Close monitoring would be required to check for abuses in the early stages of any programme that relies on retailers to distribute packs.

Notes

[1] 1.5 million packs were provided in 2000/01; the number was reduced to 1 million in 2001/02.
[2] This chapter does not explore the advantages and disadvantages of voucher systems in small projects, for instance where NGOs are working closely with project beneficiaries.
[3] Sarah Levy and Carlos Barahona (personal communication).
[4] For instance, if the programme is distributing 2 million packs, and 10% (200,000) are going missing before reaching the villages, it will only be worthwhile implementing a system of spot checks and penalties if this system a) is effective, i.e. stops the leakage of most of the 200,000 packs; and b) costs less than the packs which are 'recovered'. This is likely to be the case in a large-scale programme like Starter Pack (200,000 packs are valued at around US$1.4 million).
[5] On the concept of 'reasonable access', see Rook and Maleta (2001), pages 7–9.
[6] Dr Herschel Weeks (IFDC-Malawi), interviewed by Sarah Levy. SPLIFA retailers charged approximately five times what 2003/04 TIP distributors charged per pack (around 20 cents of a US dollar – see Chapter 3). The service provided is not exactly the same, but it is similar (receiving packs from transporters, distributing to beneficiaries, checking beneficiary vouchers and names and putting together invoices supported by vouchers). The main difference is that SPLIFA packs are larger than TIP packs, but this should have a relatively small impact on cost of distribution (size of pack would affect transport costs, but these are not included in either case).

References

Chinsinga, B., Dzimadzi, C., Chaweza, R., Kambewa, P., Kapondamgaga, P. and Mgemezulu, O. (2001) 2000/01 TIP module 4: consultations with the poor on safety nets. (Unpublished, available on CD in Back Cover Insert.)

Chinsinga, B., Dulani, B. and Kayuni, H. (2004) 2003 winter TIP evaluation qualitative study. (Unpublished, available on CD in Back Cover Insert.)

Lawson, M., Cullen, A., Sibale, B., Ligomeka, S. and Lwanda, F. (2001) 2000/01 TIP : findings of the monitoring component. (Unpublished, available on CD in Back Cover Insert.)

Levy, S. and Barahona, C. (2002) 2001/02 TIP: main report of the evaluation programme. (Unpublished, available on CD in Back Cover Insert.)

Levy, S., Nyasulu, G. and Kuyeli, J., with Barahona, C. and Garlick, C. (2002) Dedza Safety Nets Pilot Project: learning lessons about direct welfare transfers for Malawi's National Safety Nets Strategy, final report. (Unpublished, available on CD in Back Cover Insert.)

Levy, S., Nyirongo, C.C., Gondwe, H.C.Y., Mdyetseni, H.A.J., Kamanga, F.M.C.E., Msopole, R. and Barahona, C. (2004) Malawi 2003 winter TIP: a quantitative evaluation report. (Unpublished, available on CD in Back Cover Insert.)

Logistics Unit (2001) Final report: implementation of Targeted Inputs Programme

2000, Lilongwe, Malawi. (Unpublished, available on CD in Back Court Insert.)

Logistics Unit (2002) Final report: implementation of Targeted Inputs Programme 2001, Lilongwe, Malawi. (Unpublished, available on CD in Back Cover Insert.)

Logistics Unit (2003) Final report: implementation of Targeted Inputs Programme 2002, Lilongwe, Malawi. (Unpublished, available on CD in Back Cover Insert.)

Nyirongo, C.C., Msika, F.B.M., Mdyetseni, H.A.J., Kamanga, F.M.C.E. and Levy, S. (2003) Food production and security in rural Malawi (pre-harvest survey). Final report on evaluation module 1 of the 2002/03 Extended Targeted Inputs Programme (ETIP). (Unpublished, available on CD in Back Cover Insert.)

Rook, J. and Maleta, M. (2001) Malawi National Safety Nets Programme: private sector participation in the delivery of safety net transfers: issues & mechanisms. MTL consult. (Unpublished)

II Methodology of the Evaluation Programme

5 Design of the Evaluation Programme

Ian M. Wilson

5.1 Introduction

This chapter provides an account of the thinking behind the design of the evaluation programme for Starter Pack. A programme evaluation in a developing country may sometimes be criticized as a one-point-in-time, donor-defined, foreign-run exercise. It may also be seen as having a tangle of objectives and limited awareness of and relevance to local issues and sensitivities. We sought to minimize these problems by using a design based on 'modules'. This chapter describes how such a modular design helped to overcome some of these problems. In particular, it allowed us to involve local consultants in the evaluation process and to utilize the programme to help upgrade their skills.

The evaluation programme design described in this chapter had a number of advantages: it had sensible amounts of time and resources and a relatively long period of continuity of organization, as well as building on a precursor activity. The main evaluation scheme had its beginnings in 1999, after the first round of Starter Pack seed and fertilizer distribution had been completed and evaluated, when it was becoming clear that the free inputs programmes would continue, and that the evaluation would have to address a wide variety of searching questions. There had been a resource-constrained evaluation of the 1998/99 Starter Pack, which – unsurprisingly – had been unable to respond to all the challenges thrown at it by interest groups.

5.2 Design Problems and Challenges

This author was originally invited to Malawi to suggest an alternative for the 1999/2000 Starter Pack evaluation, and was given the opportunity to talk to a number of stakeholders. A new evaluation strategy was required. There were clearly several limitations that prevented it from following textbook models:

- Detailed and reliable census data were not available. If they had been, they could and should have been used to refine the selection of representative samples.
- In 1999/2000, Starter Pack was aiming at universality rather than targeting (all smallholder households were to receive a pack), so comparisons with a more or less equivalent group of non-beneficiaries were not on the cards. This contrasts with programmes where there are, for example, 'with' and 'without' districts.
- Comparisons on a 'before and after' basis were limited by the late start to evaluation and by annual last-minute decisions about the size and coverage of the programme (leaving insufficient time for conducting 'baseline' studies). In addition, it would have been difficult to argue that the 'before' situation represented a 'normal' without-intervention situation, given large year-to-year variations in weather patterns and other factors affecting farming.
- Human and financial resources were limited, so it was not feasible to consider, for instance, very large sample, time-consuming surveys.

The approach to evaluating Starter Pack also needed to be flexible to reflect changes in the programme and in the focus of interest of stakeholders over time. The breadth and depth of interest of the research are now attested by the contents of this book, by the evaluation reports and related data available in the CD archive of the evaluation programme (SSC, 2004; see Back Cover Insert) and by a barrage of Malawian newspaper articles. An approach was needed that allowed a wide range of issues and questions to be addressed sensibly.

First acquaintance with the evaluation challenge revealed enough of the complexity of stakeholder interests to make it clear that any unified evaluation exercise would be too complicated to administer. There was a need to address a number of very different facets of Starter Pack, and later of the Targeted Inputs Programme (TIP). For example, some evidence was clearly needed on whether the free seed and fertilizer distribution exercise was providing agronomic returns, in particular the contribution to national maize production (see Box 5.1). However, there was also fierce argument about other concerns such as: How many rural households should be included in the programme? Should farmers receive the packs completely free, or would this encourage a 'culture of dependency'? Should packs be provided only to those who are prepared to pay for them either with cash or with labour, working on a public works scheme? Was the distribution of hybrid maize seed and chemical fertilizer undermining efforts to promote sustainable agriculture? Many of these concerns could not be addressed by a survey. Instead, a range of research methods, including qualitative approaches, was needed.

Another area of concern was the role of external consultants. The evaluation of the 1998/99 Starter Pack was criticized, for example in the Ministry of Agriculture (MoA), because although local resources had been utilized for

> **Box 5.1.** Estimating maize production.
>
> A core question was how to estimate maize production. The 1998/99 Starter Pack evaluation (Longley et al., 1999) had attempted a crop measurement approach based on crop-cutting. This involved identifying Starter Pack plots planted as per instructions on 0.1 ha, and measuring their maize yields. These were to be compared with the same farmers' maize yields, derived in the same way, from plots planted with their own seed. There turned out to be innumerable variations and complications in how the Starter Pack seed and fertilizer were used, and the comparison was finally based on small subsamples of (perhaps untypical) farmers who followed instructions.
>
> The 1999/2000 Starter Pack evaluation adopted a different approach, based on the view that an attempt to measure incremental yield attributable to Starter Pack under complex field conditions would not be accurate enough to merit the high cost of the exercise.
>
> Instead, we decided to attempt to obtain farmer estimates of maize production for 1998/99 and estimates/forecasts of production for 1999/2000.[1] We would not aim for high levels of precision, but would look for trends and orders of magnitude. Owing to the large number of respondents in our samples, we would be able to analyse the variations in the size of the effects observed in relation to farmer characteristics such as wealth/poverty, sex of household head and location. At the centre of this new approach was the idea that we would triangulate the results from different methods (see Chapter 6).
>
> Source: Adapted from Levy et al., 2000.

data collection, the analysis and report-writing had been done largely by international consultants. There were valid reasons why the financial sponsor, the UK Department for International Development (DFID), was propelled in this direction: the lead time for a second Starter Pack programme was short and therefore answers about the first Starter Pack were needed quickly; and the report had to pass muster quickly with funding agencies outside Malawi. There appeared to be very few working groups in Malawi which had sufficiently well-established track records in managing all aspects of such a pressurized project.

There were, all the same, compelling reasons why local knowledge needed to be brought to bear on the issues raised by Starter Pack and why a variety of local experts needed to be involved. It was a large programme, with impacts in every corner of the country. It had substantial implications in areas associated with a range of disciplines including agriculture, economics, sociology and politics. The study of this important intervention was a legitimate interest of groups involved in research, policy, logistics, agricultural extension and many other fields. If there was an evaluation exercise that gave no voice to such groups, it could be expected to raise resentment, opposition and disengagement.

One particular difficulty in involving local expertise lay in the breadth of the set of issues to be studied. There were some Malawian individuals and groups with specific experience in one sector or field, but not in others. It appeared that Starter Pack would require a consortia of different expertise to tackle all the issues. It also appeared that there were relatively few groups which could demonstrate that they had enough experience of management to operate to the standards of general management and financial reporting and control that are required of international consultants.

Another difficulty was that some of the issues that were suggested as suitable topics for evaluation would require quantitative data collection (surveys), of a high standard and in a tight time frame, whereas others were more concerned with developing social and attitudinal understandings, best researched using qualitative approaches. The qualitative enquiries needed in-depth understanding, but at the same time had to meet the same criteria as the quantitative themes: they needed to produce results that were meaningful and reasonably representative at the national level because Starter Pack was a nationwide programme. As Malawi encompasses important variations in farming situations and social structures, qualitative and quantitative research exercises alike would require careful attention to issues of respondent sample selection, data collection instrument validation, data entry and management quality and the selection of suitable data summaries in analysis (see Chapter 6).

5.3 The Modular Structure

How should an evaluation programme be structured in these circumstances? It needed to include or engage with a very wide range of legitimate interests and provide insights into a dauntingly diverse range of issues. Handling all the issues in one large study would be inefficient, because each topic – and the research method which was most appropriate for it – would require a different approach to sampling, different professional expertise and different amounts of time spent in each site. Putting all the issues together would mean working on the basis of the largest common denominator, and much time and resources would be wasted as people waited for each other to do different parts of the study. Moreover, in the villages visited, farmers would be asked to spend too much of their valuable time answering questions from different experts.

The proposal discussed by this author and the DFID project manager was based on a series of separate studies or 'modules'. We argued that it would be desirable that:

1. Different facets of the evaluation be administered as separate exercises, because they required different knowledge and skills. Those who could effectively measure agricultural production and the contribution of Starter Pack seed and fertilizer needed specific knowledge of primary agricultural data collection and they also needed to have quantitative analysis skills. Different people would be better placed to conduct research on demographics, or to run studies that required consultations with communities, using participatory methods.

2. The separate exercises be contracted out to different groups, so that each had a manageable task to do and was not over-stretched. Each could then have well-defined and tightly specified objectives. A further benefit of dispersed contracting would be that any difficulty that seriously affected one group would not undermine or cripple the whole evaluation programme. This, of course, raised some difficulties: there could be expected to be some disparities of standard; at the interfaces between the themes there might be issues left uncovered, or there might be some overlap and even some contradictory signals; there could be expected to be some weaker reports. In the event, these difficulties were not felt to outweigh the advantages of modularization.

3. Each of the exercises should have its own methodological approach, appropriate to the objectives of the study. The samples should be of different sizes, reflecting the particular requirements of the approach, and should be mutually exclusive to avoid overburdening respondents/participants in any one village.

5.4 Capacity-building and Ownership of Findings

How could the evaluation programme increase the involvement of Malawian researchers? It appeared that a good proportion of those whose skills might contribute most might not qualify: they would not be likely to meet all the necessary criteria to run an evaluation that would meet international standards. Therefore, it was proposed that a secondary purpose of the evaluation exercise should be to assist Malawian consultants in their professional development. The goal was that Malawian groups should be given the opportunity to practise working in the same way as international contractors, but technical support would be provided by external managers who could diagnose the Malawian contracting parties' key limitations, and who could provide supervision, training and advice. The Statistical Services Centre (SSC) of The University of Reading was contracted to provide this technical support, in particular to provide:

- Coordination of sampling strategies of the modules to ensure representative results and generalizable conclusions;
- Quality control for fieldwork, information management and analysis;
- Help to the module teams in structuring and writing final reports; and
- Facilitation of an end-of-programme results seminar.

The preparation for the results seminar was a good example of the managed process. Up to this point, each team was working hard on its own themes without much interaction. Before making their presentations to a large national and international audience, the teams were brought together for 2 days of preparatory work, where they listened to each other's presentations and acquired familiarity with the overall picture of the evaluation and its implications. Presentations were intensively critiqued, edited and checked, so that in the main presentation the audience saw a clear, lively, relevant distillation of the reports.

In the first instance, the advice and help provided by the external managers had to compensate for any diagnosed limitations. Over the longer term, it was intended to assist the teams towards greater self-sufficiency in handling evaluation contracts. The latter objective was not fulfilled where teams of local consultants were involved in the evaluation for 1 year only, but proved successful with those teams which won contracts several years running, over the 4-year period of the evaluation programme (1999–2003).

It could be argued that this approach involved a trade-off: if extra resources went into capacity-building rather than directly into results evaluation, this would be a diversion from the main purpose. The defence against this argument was that the involvement of, and therefore the assistance to, the Malawian partners was necessary if the evaluation was to have the benefit of their local knowledge and to engage with Malawian thinking about the issues involved.

Ownership of the evaluation was considered and it was concluded that a results seminar would be desirable not only to present the substantive findings to government, donors and the wider public but also to provide a public platform where the achievements of the Malawian teams could be showcased. After the results seminar, reports from the evaluation studies were made available on paper where needed, but the main method of dissemination was on CD, as this technology was becoming readily and cheaply accessible.

Involvement with the Starter Pack and TIP evaluation programme was rewarding for most people because of the high degree of public interest in the issues it raised. From an early stage, it was evident that there had to be a well-organized process to define the themes that were to be studied. Naturally, this involved stakeholder consultation, which identified and considered the issues. However, a weakness of the evaluation programme in later years was that there was insufficient time allocated to stakeholder consultation, and, as a result, some members of the donor community questioned the objectiveness of the results, which were perceived as being influenced by DFID's policy agenda. This perception was unfounded, in the author's opinion, as all members of the evaluation team took their professional responsibilities as independent evaluators seriously. However, it was understandable. An important lesson is that where the programme under evaluation is a controversial one, as in the case of Starter Pack, all stakeholders need to be consulted on a regular basis.

5.5 Conclusion

The evaluation process sketched in this chapter has borne fruit: the evidence presented in Part 3 of this book is a result of the work it stimulated. The modular design and the sincere attempt to involve local evaluators have certainly provided a number of Malawian groups with enhanced experience of working to the standards of an internationally funded contract, at the same time offering capacity-building opportunities. The involvement of their institutions, and many other parties, in results presentation seminars has stimulated considerable interest in the dissemination and discussion of the findings. No one would claim that the exercises undertaken were flawless, or that the policy impact always matched the hopes of those who carried out the research, but with the availability of this book and of the detailed evaluation archive contained in its Back Cover Insert, readers can judge the merits of the scheme for themselves. We encourage others to take forward this effort.

Note

[1] The survey was to be carried out at around the time of the maize harvest but before the harvest was over, so production would have to be forecast in many cases.

References

Levy, S., Barahona, C.E. and Wilson, I.M. (2000) 1999/2000 Starter Pack evaluation programme main report. Statistical Services Centre, The University of Reading, UK. (Unpublished, available on CD in Back Cover Insert.)

Longley, C., Coulter, J. and Thompson, R. (1999) Malawi rural livelihoods Starter Pack scheme 1998/99: evaluation report. Overseas Development Institute, London. (Unpublished, available on CD in Back Cover Insert.)

SSC (2004) The monitoring and evaluation archive of Malawi's Starter Pack and Targeted Inputs Programmes 1999-2003. Statistical Services Centre, University of Reading, UK. (Available in Back Cover Insert.)

6 Experience and Innovation: How the Research Methods Evolved

CARLOS BARAHONA

6.1 Introduction

In 1999, UK-based consultants from the Statistical Services Centre (SSC) of The University of Reading and Calibre Consultants were contracted by the Department for International Development (DFID)-Malawi to manage a Monitoring and Evaluation (M&E) programme for Starter Pack. The M&E exercise was conceived as an independent, impartial evidence-gathering operation to provide evidence to decision makers on the direct and indirect effects of the Starter Pack. The main emphasis was on evaluating the impact of Starter Pack and later of the Targeted Inputs Programme (TIP), but in 2000/01 the programme also included a Monitoring Component.

From 1999 to 2003, the DFID office in Malawi contracted the management of the Starter Pack/TIP M&E programme to the SSC and Calibre Consultants on a year to year basis. The role of the managers was to consult stakeholders on the research agenda and ensure that the programme produced a reliable and coherent body of information. In practice, it involved defining terms of reference for each study, selecting teams of local consultants to conduct the studies, provision of methodological support to each team, quality control to ensure that the work achieved international standards, critiquing and helping to improve the final reports of the study teams, and the provision of annual summaries of the research findings to decision makers. The managers also ensured strict financial control of each study and made an explicit link between delivery of products and payments to consultants (see Chapter 7).

6.2 The Structure of the Programme

The M&E programme was designed using a modular approach (see Chapter 5). Each module corresponds to an independent study, and the modules are inter-related. This structure allowed the programme to incorporate statistical rigour, triangulation and flexible adoption of approaches and methods. Integration of the findings of the studies occurred at 'results seminars' (see Chapter 5) and in summary reports and policy briefings prepared by the managers of the programme. The modules included:

- surveys with national coverage for which results are representative of the population of rural smallholders in Malawi;
- studies that used participatory methods for exploring complex questions (also designed to generate information that is representative of rural Malawi); and
- case studies that do not claim to be representative of the whole of rural Malawi but provide insights into complex issues.

The set of studies carried out by the Starter Pack and TIP M&E teams is shown in Table 6.1. The relationships between modules can be described as:

Modules sharing at least one objective. In these cases, related modules addressed the shared objective independently. Sometimes the same methodology was used. For example, when assessing whether TIP benefited the poorest, information that allowed the construction of a poverty index (see Section 6.5.2) was collected by each module and the results were analysed following the same methodology. After a period of testing and development, this poverty index was adopted by all the quantitative modules of the evaluation and comparison of results across modules was possible. In other cases, different methodologies were employed to explore the same issues. For instance, when assessing farmers' links with markets in 2000/01, Module 2 Part 1 used a quantitative methodology based on a survey and Module 2 Part 2 based its approach on case studies.

Addressing the same subject with a different degree of detail. Survey-based modules asked simple questions about a particular topic, but had wide coverage in terms of sites and number of respondents. Other modules were able to study the same issue in greater depth; but because of the more complex analysis that took place, these modules had restrictions on coverage and sample size. For example, the food production and security surveys enquired about receipt and condition of the agricultural extension leaflet; they allowed us to estimate percentages of leaflets that arrived in good condition and percentages of farmers following the recommended practices. Two other modules were commissioned, in 2000/01 and 2001/02, to look at the TIP extension effort, assess farmers' views about the messages and consider ways of improving the leaflet. These two modules relied on the use of participatory methods (in 27 villages in 2000/01 and 21 villages in 2001/02). Their design incorporated statistical principles in such a way as to make it possible to claim that the results are representative (see Section 6.4). The integration of the food production and security survey results and the modules using participatory methods helped to provide a complete picture of the agricultural extension issues (see Chapter 13).

Table 6.1. Modular structure of the M&E programme.

1999/2000	2000/01	2001/02	Winter 2002	2002/03	Winter 2003
Module 1: Agronomic Survey	Module 1: Food Production and Security	Module 1: Food Production and Security in Rural Malawi (pre-harvest survey)	Winter TIP Evaluation Survey	Food Production and Security in Rural Malawi 2002/03	A Quantitative Evaluation of the 2003 Winter TIP (survey)
Module 2: Micro-economic Impact and Willingness to Pay	Module 2.1: A Quantitative Study of Markets and Livelihood Security	Module 1: Food Production and Security in Rural Malawi (post-harvest survey)			A Qualitative Evaluation of the 2003 Winter TIP
Module 3: Gender and Intra-household Distribution	Module 2.2: A Qualitative Study of Markets and Livelihood Security	Module 2: TIP Messages: Beneficiary Selection and Community Targeting, Agricultural Extension and Health (TB & HIV/AIDS)			
Module 4: The Impact of Starter Pack on Sustainable Agriculture in Malawi	Module 3: Agricultural Communications and the HIV/AIDS Leaflets				
Module 5: Measuring the Size of the Rural Population in Malawi	Module 4: Consultations with the Poor on Safety Nets				
	Monitoring Component				

The modular approach allowed the evaluation to achieve:

- The integration of findings from surveys, from studies based on participatory methods with nationwide coverage and from in-depth case studies in purposively selected villages.
- Triangulation of findings about issues that were deliberately included in more than one study. Since each module conducted its work independently and the design of the whole programme included overlapping of specific issues, key findings could be compared and validated.
- Maximum geographical coverage, while minimizing demands on farmers' time.
- Efficient interaction between Malawian consultants and the management team.
- Establishment of appropriate quality control systems in each module, which contributed towards high-quality results.

6.3 The Evolution of the Programme

The evolution of the M&E programme was affected by the questions that decision makers had about Starter Pack and the TIP. The number of questions at the beginning of the programme was large, leading to the need for five studies in the 1999/2000 season and six in 2000/01. By 2002/03, the M&E programme was limited to a single study, as most of the key questions had been researched by this time.

6.3.1 The surveys

One module ran every year: the food production and security survey. This survey had national coverage, interviewed around 3000 households and provided information on the impact of Starter Pack and the TIP on agricultural production, use of the programme inputs, access to inputs from sources other than Starter Pack/TIP, poverty profiles of beneficiaries and non-beneficiaries and household food security status. The choice of a survey as the tool to tackle these issues was determined by the need to produce quantitative information about them, but imposed a constraint on the complexity of the questions that could be asked.

The series of data from the evaluation surveys provides a useful tool to analyse the impact of the free inputs programmes and of the changes that were made to them. It also serves as a reliable source of information on trends in maize production and food security in the smallholder sector.

In 2001, when the results from two surveys (2000/01 Modules 1 and 2.1) were checked, the evaluation team made a series of visits to policy makers and donors to discuss the finding that maize production had fallen sharply with respect to the previous year. These results were presented in writing in September 2001 (Levy and Barahona, 2001). They constituted the earliest independent warning of the 2001/02 food crisis. This was possible because the survey data allowed us to make direct comparisons between maize production on a year to year basis, and because of the national coverage of the surveys and the relatively quick processing of the data. The quality control system established for the work (see Section 6.6.1) meant that we had a high level of confidence in the findings.

During the 2001/02 food crisis and the following year, special attention was paid to information collected about use of coping strategies, in order to measure food security at household level (see Section 6.5.1). The data allowed the evaluation teams to classify households into three levels of food security: food secure, food insecure and extremely food insecure. Figure 6.1 shows the percentage of households that were extremely food insecure in each month of 2001/02 and 2002/03 – highlighting the timing and severity of the 'hungry period'.

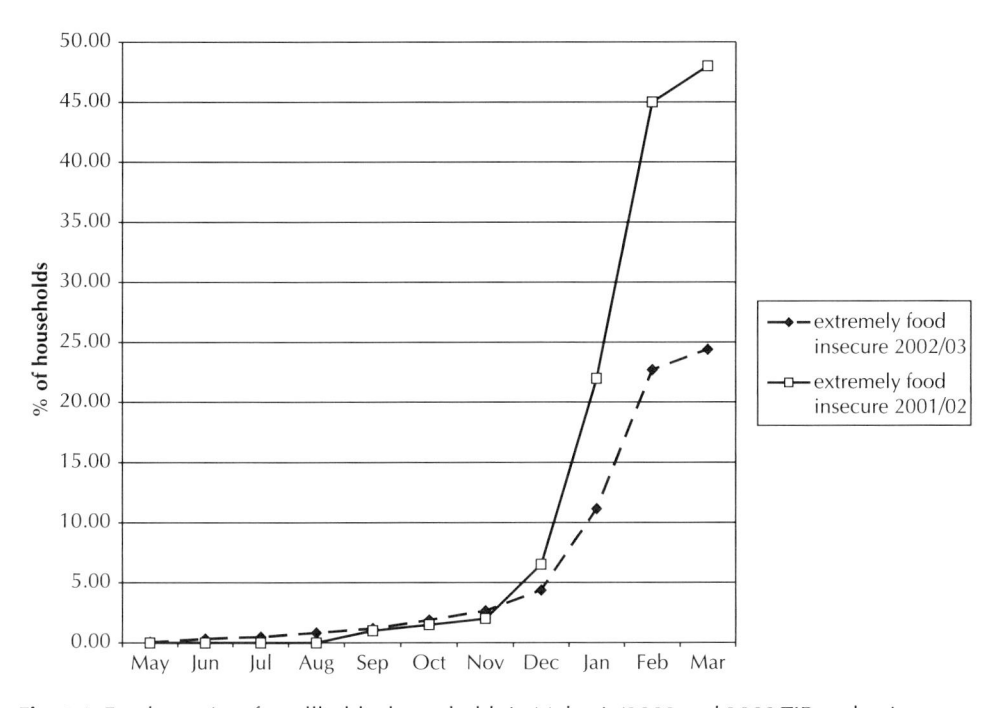

Fig. 6.1. Food security of smallholder households in Malawi. (2002 and 2003 TIP evaluation surveys.)

6.3.2 The 'qualitative' studies

More complex questions were also asked by decision makers. In some cases, these were addressed as a one-off study, either because this was enough to answer the question or because the question went down the list of stakeholder priorities. Some examples are:

- What is Starter Pack's impact on sustainable agriculture? (Module 4, 1999/2000).

- Were there as many rural smallholder households as was claimed by the register of beneficiaries for Starter Pack? (Module 5, 1999/2000).
- What level of integration with markets do rural households have? What is the role of maize in rural livelihoods and the nature of *ganyu*? (Module 2, 2000/01).

In other cases, two studies examined complex issues. For example:

- After the introduction of TIP, it was considered important to find out whether the community targeting mechanism used had succeeded in targeting the poorest households. What was the perception of communities about this method of selecting beneficiaries? Are there any criteria that communities consider acceptable for selecting beneficiaries? Is there an optimum programme size for achieving a successful targeting outcome? (Module 4, 2000/01; and Module 2, 2001/02).
- An assessment of agricultural and health messages accompanying the free inputs programmes. In relation to the agricultural technology message, questions included the suitability of the materials distributed, effective mechanisms for conveying the information and a general assessment of agricultural extension in relation to the TIP. Regarding the health messages (especially on HIV/AIDS), there were concerns about their cultural acceptability, how they were received by people of different ages and sex, and what health information is perceived as useful and important in rural communities (Module 3, 2000/01; and Module 2, 2001/02).

6.4 Participatory Methods

Complex questions demand the use of sophisticated methods of enquiry such as participatory methods, but at the same time the results need to be capable of informing national policy. Traditional participatory approaches do not generate results that can claim to be representative of a population other than in the sites that are visited. This posed a methodological challenge that led the M&E team to develop study designs integrating participatory approaches with statistical principles.[1] So that the findings would be representative of the population of smallholder farmers in Malawi, the study designs included the following elements:

1. Working in a larger number of sites than is common for most studies that use participatory methods.
2. Selecting sites using statistical sampling methods. The M&E programme benefited from having a reliable list of villages in Malawi provided by the Logistics Unit's beneficiary register,[2] which was used as a sampling frame.
3. Developing, testing and improving a detailed methodology for the work to be carried out within each village; and production of a field manual which standardized the approach, so that key information was collected in the same way in all sites (see Section 6.4.1).

4. Using a uniform format for reporting the findings from each site. This was called the 'debriefing document', and was a key element in the integration of results from the field work.
5. Training a field team in the use of the field manual and the debriefing document, as well as facilitation and note-taking skills.
6. Producing an integrated analysis of the findings. This included the use of statistical methods of analysis for numerical information and traditional social sciences methods for non-quantifiable information.

6.4.1 Statistical principles

There are three aspects that make the use of participatory methods in the M&E of Starter Pack and TIP different from traditional Participatory Rural Appraisal (PRA), and that make it possible to consider the results as representative.[3] These are:

- *Standardization*. The fact that the methodology was applied in the same way in every site visited allows us to compare findings across sites. In standardizing the methodology, there is a careful balance to strike: while the findings need to be comparable across sites, the standardization process must endeavour to keep the flexibility of participatory tools. This implies that standardization applies to the researcher's approach and behaviour, while allowing local participants to express their views and concerns.
- *Structure*. This is an aspect that has been borrowed directly from statistical approaches and is reflected in the design of the study, especially the sampling scheme. First, it is important to understand the relationships between the units of interest to the research (about which information is required). An example of relationships between units can be seen if we consider that households are normally 'organized' in hierarchies such as villages (clusters of households) and districts (groupings of villages). At the household level, units can be subdivided into groups, for example food secure, food insecure and extremely food insecure households. Second, we need to define the relationships between the units of the enquiry and its objectives, in order to know what type of information should be collected from whom, and what possible explanatory factors should be taken into account.
- *Coverage*. This is directly related to whether the result can be considered representative of a population. When we claim that 'results are representative' there are a series of conditions that need to be satisfied, including three relating to coverage: (i) a population of interest must be defined prior to conducting the research; (ii) the selection of the units about which the information is collected has to give every unit in the population a fair chance of being included in the study; and (iii) the number of units in the study has to be large enough to capture the variability of the information of interest and the diversity of units. The M&E programme for Starter Pack and TIP endeavoured to fulfil these three conditions in the studies

that used participatory approaches. In all cases, sampling was based on the best available list of villages in the country and allowed a fair, measurable probability of inclusion to every village; and efforts were made to visit the largest possible number of villages within the resource constraints. In relation to the number of sites, it must be stressed that although the number of villages visited varied between 20 and 40, numerical information was generally collected at household level, so the number of units on which information was collected was large and comparable with standard survey methods.

6.5 Indicators and Tools

The selection of the methodologies and tools used in the M&E programme was done on the basis of the questions asked by decision makers. This section presents the more innovative of the methods used.

6.5.1 The food security indicator

A household food security indicator was developed by the survey modules (Sibale *et al.*, 2001; Nyirongo *et al.*, 2002, 2003). This approach looks at symptoms of food insecurity – coping strategies adopted by households in times of food scarcity – rather than availability of food from different sources, on the basis that people will only resort to coping activities when they cannot access food from any source. Thus, the indicator captures availability (or lack) of food from the household's own production *and* from off-farm sources.

The evaluation chose the following coping strategies to build the food security indicator:

Moderate coping strategies:

- Eating *nsima* from green maize
- Eating maize bran
- Eating only fruits (no *nsima*)
- Eating vegetables only (no *nsima*)
- Eating sugarcane only (no *nsima*)

Extreme coping strategies:

- Eating *nsima* from maize cobs
- Eating only wild roots and tubers (no *nsima*)
- Eating only wild fruit, mushrooms etc (no *nsima*)
- Eating nothing for the whole day

Households were asked just before the maize harvest (when the surveys were carried out), whether they had resorted to any of these strategies since the previous maize harvest. If they said yes to any of the strategies mentioned, we asked them in which month they started to use it. We then classified the households interviewed as food secure (not using any coping strategies), moderately food insecure and extremely food insecure. In some module reports, an alternative approach to defining moderate or extreme food insecurity was used, based on whether households began to resort to coping strategies before December.

This indicator is capable of documenting the marked seasonal pattern of food insecurity in rural Malawi. The M&E programme reports pointed out that the number of households facing food insecurity in rural Malawi varies sharply with the time of year (see Fig. 6.1).

6.5.1.1 Early warning
This food security indicator cannot act as an early warning system. Monitoring it would not provide information in time to allow planning and implementation of relief interventions. This is because the symptoms of extreme food insecurity are not sufficiently in evidence to be able to detect them until well into the 'hungry period'.

The best data for early warning purposes in Malawi are the quantities of main staple foods produced each year (in particular maize) and trends in the prices of maize, cassava, rice and sweet potatoes.[4] Tracking food prices during the early months of the 'hungry period' (September–November) is normally sufficient to predict food crises in the latter part of the hungry period (December–March). This can be complemented by monitoring of the proportion of households buying (or trying to buy) food in the early months of the 'hungry period', indicating that they have run out of own-produced food sources. It is also important to take into account forecasts of food imports (see Chapter 18).

6.5.2 The poverty index

The need to determine whether the benefits were reaching the poorest in the TIPs led to the construction of a poverty index for use in the surveys. The index is derived from a combination of assets and indicators of income of the household and provides a practical tool to distinguish broad poverty levels among the sampled households.

The index is built by creating two intermediate indices: (i) an assets index calculated from a weighted sum of the relative value of assets owned by the household, including small animals, cattle, oxcarts, radios and bicycles, and (ii) an income index based on household income over the month previous to the interview for a range of regular sources of income and income from agriculture-related activities. The income index does not attempt to measure income as such; it only attempts to place households into broad income categories. The limits for the categories of the income index were changed from year to year to reflect rural inflation. The two intermediate indices are finally combined into a new 'poverty index' (see Table 6.2) with five poverty categories – from poverty category 1 (poorest households) to poverty category 5 (least poor).[5]

Table 6.2. Derivation of poverty index for 2002/03.

Income index	Assets index			
	Less than 2	2 to 30	31 to 70	71 or higher
Up to MK 171	1	1	2	3
MK 172 to 514	1	2	3	4
MK 515 to 1372	2	3	4	5
MK 1373 plus	3	4	5	5

Note: The scale for assets index is derived from relative market values for the assets included in the index.
Source: Nyirongo et al. (2003).

Nyirongo *et al.* (2003) describe the poverty index as a relative measure of poverty. It 'can be used for comparing relationships between poverty and other variables or for assessing 'poverty profiles' of beneficiaries and non-beneficiaries of TIP, food aid, etc. It is *not* an absolute measure of poverty like the poverty line calculated by the Integrated Household Survey (IHS)'. The usefulness of this index is restricted to comparisons between households that are visited in a single study, or studies that are run simultaneously. This was not a problem for the M&E programme, as this was our only objective. We could not have attempted to make comparisons over time without good information on the evolution of real household incomes in rural Malawi – which was beyond the scope of the Starter Pack/TIP evaluation surveys.

An example of the use of the index is the comparison of poverty profiles. Figure 6.2 shows that the poverty profiles of beneficiaries and non-beneficiaries were very similar in 2001/02; in other words, there were similar percentages of beneficiaries and non-beneficiaries in each poverty category, from the poorest to the least poor (wealthiest). This shows that poverty targeting did not succeed in 2001/02. However, from the point of view of the evaluation, the similarity in poverty profiles between benefici-

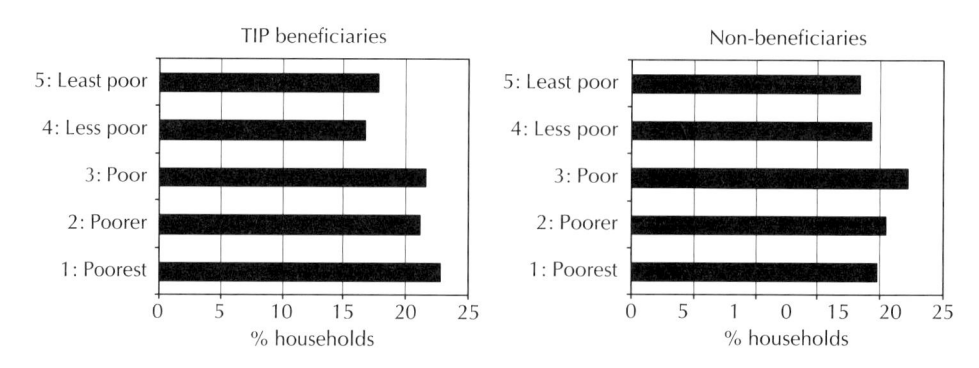

Fig. 6.2. Poverty profiles of TIP beneficiaries and non-beneficiaries in 2001/02. (Adapted from Nyirongo *et al.*, 2002.)

aries and non-beneficiaries proved helpful, as we were able to use non-beneficiaries as a comparison group[6] in order to calculate the TIP's contribution to maize production.

6.5.3 Estimating the total number of households in rural Malawi

The question about what is the number of smallholder households in rural Malawi was crucial for the planning and execution of Starter Pack and the TIP. In 1999/2000, the first list of beneficiary households for Starter Pack compiled by the Ministry of Agriculture (MoA) contained 3.7 million names (Wingfield Digby, 2000). Decision makers were concerned that this number was too large, as the last census in 1998 had reported a total rural population of 8.5 million (1.95 million households). The number of beneficiaries on the Starter Pack Logistics Unit (SPLU) register was eventually pared down to 2.86 million.

In November 1999, the M&E programme was asked to find out whether the number of beneficiaries on the SPLU register matched the number of smallholder households on the ground. Discrepancies between the numbers coming from the register and census numbers could have been due to several reasons:

1. Starter Pack aimed to distribute one pack per 'farm family', and a farm family might have been different from the census definition of a household;[7]
2. There may have been undercounting in the census figures, as was rumoured in Malawi at the time the census was conducted;
3. There were incentives for households to register more than one beneficiary for Starter Pack; and/or
4. Corrupt local officials may have inflated the number of beneficiaries for Starter Pack in order to obtain packs for themselves.

The 1999/2000 Module 5 study 'Measuring the size of the rural population in Malawi' (Wingfield Digby, 2000) was designed to address this issue. It used a combination of participatory tools and statistical methods, and carried out a Ground Truth Investigation Study (GTIS) in 54 villages. Barahona and Levy (2003) describe it as follows:

> Through the [participatory] mapping and marking of households, the study produced a 'participatory household listing'. A total of 6326 households were listed in the 54 villages studied. Each household was given a unique number, which could be used to identify it. The fact that the household listing was done by a group using a visual tool (the map) contributed to its reliability, as each of the participants was able to check that the information being recorded was correct...
>
> The researchers then visited every household and interviewed the household head using a two-page questionnaire identified with the same number as on the map... The main purpose of this questionnaire was to collect data on the number of people in each household. Additional demographic data and information about which members of the household cultivated some land, registration for Starter Pack and receipt of packs were also collected.

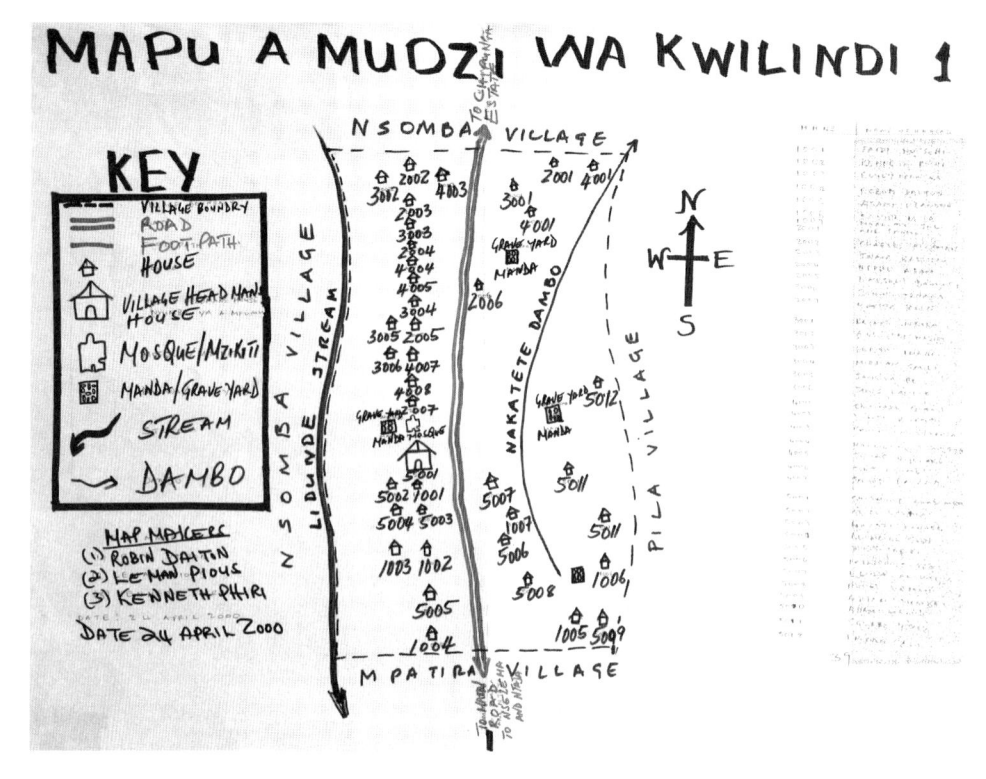

Fig. 6.3. Participatory map for Module 5. (GTIS study for Module 5, 1999/2000 Starter Pack evaluation.)

P. Wingfield Digby, a former UK government statistician with experience in handling demographic data in Africa, was then able to provide a national population estimate by scaling up the data for the sample. The analysis of the data produced unexpected results. The total number of households was estimated at 2.78 million and the rural population was estimated at 11.52 million. In the villages visited, the data showed a fair agreement between the total number of beneficiaries ('farm families') eventually registered by the SPLU – after the register was pared down to 2.86 million – and the number of households on the ground according to the GTIS (see Fig. 6.4).

These results provided evidence to ask a further question: Are the figures from the 1998 census a serious undercount of the population of Malawi? This question was beyond the scope of the M&E programme and was passed on to the National Statistical Office (NSO). To date the NSO has not provided an answer.

6.5.4 Community mapping with cards

The participatory mapping technique used by Module 5 (1999/2000) was developed during the TIP evaluations to become a tool for collecting reliable

Research Method Evolution 89

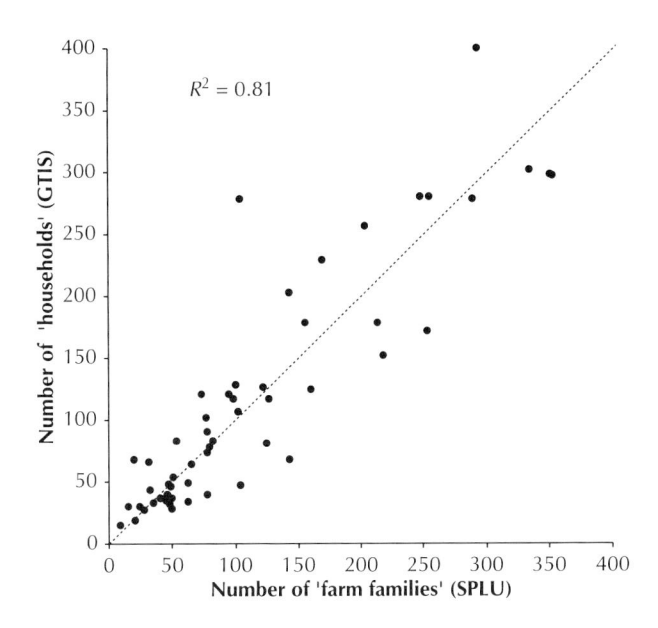

Fig. 6.4. Relationship between the number of households in villages and the number of registered beneficiaries ('farm families'). (Wingfield Digby, 2000.)

[Scatter plot: x-axis "Number of 'farm families' (SPLU)" from 0 to 400; y-axis "Number of 'households' (GTIS)" from 0 to 400; $R^2 = 0.81$]

numerical information using participatory approaches (Barahona and Levy, 2003). The procedure was to ask five to ten community members to draw a map of the village. The participants were asked to mark every household in the village on the map and to give it a number. Then they prepared a card for each household, with the name of the household head and the household number as shown on the map. It was vital that every household in the village appeared on the map and had a card with the same number as on the map.

The cards were a useful visual aid for stimulating discussion and could be used to collect data about key characteristics of households, such as food security status, whether or not they had received a pack of free inputs, and whether they met the criteria to qualify as a beneficiary of the programme. Some of the information on the cards could be processed with the participants to produce village-level numerical analyses such as tables showing whether the 'right' households were receiving the packs. Later, the researchers could process the data from all villages and produce national data summaries.

Community mapping with cards proved to be a very useful technique for generating numerical information through participatory methods, and was used in several evaluation modules (see Chapter 11). It has the advantage that collecting information on each household and recording that information on one card per household – not in aggregated form as similar participatory activities do when they pile or group households – is equivalent to a full census of the village. This eliminates sampling error within sites and allows the use of statistical methods for analysis. Two additional advantages of this technique are that it is simple to do and can be understood by most participants, so there is a good chance of producing reliable results; and household-level information that is in the public domain can be collected efficiently by asking participants to act as key informants.

6.6 Standards and Ethics

6.6.1 Quality control

The management of the M&E programme placed a strong emphasis on ensuring the quality of the data collected. This was achieved through the establishment and refinement of quality control checks designed to minimize errors. Three modules were rejected on the basis of not having achieved the standard required by the M&E programme.

Quality control systems for the evaluation surveys were based on best practice guidelines for this type of work. They included development of questionnaires with local counterparts (some sections derived from consultations with farmers), translation into the two main local languages – Chichewa and Tumbuka, field testing, training of interviewers including practical field exercises, supervisor guides for quality checks in the field including special consistency checking forms to be completed for each questionnaire, a mechanism of penalties and incentives to reduce field-level errors, data entry forms developed to minimize errors, double data entry of all questionnaires and double checking of reports and statistical analysis by the managers of the M&E programme.

For modules based on participatory approaches, the system included development of tools with local counterparts, extensive field testing, development of field manuals and debriefing documents, thorough training of researchers and research assistants, analysis workshops and critical reading of the final reports by the managers of the evaluation and independent experts brought in to help review the results.

Each module report contains a detailed description of the methodology and an assessment of areas where problems were experienced. In order to promote transparency, the results from all the modules of the M&E programme were disseminated in an electronic evaluation archive available on CD (SSC, 2004; see Back Cover Insert). In addition to the reports, the CD includes the complete set of survey questionnaires, raw data files, the results of participatory activities, and field manuals and completed debriefing documents.

6.6.2 Ethical issues

The evaluation team had to deal with a series of ethical issues in the course of the research. Of particular interest are those associated with the integration of participatory approaches and statistical methods. Perhaps the most important decision in this area was taken in 2002, when we decided to stop running studies based on participatory approaches after it became evident that decision makers were ignoring communities' views. The participatory work depended on research teams being able to establish agreements with participants that they would get communities' views heard by decision makers, so that appropriate decisions could be taken on the future of the intervention. Once it became impossible to make such a commitment in good faith, we could not continue with this type of work.

Other ethical issues that arose had to do with: the need to ensure transparency and consent about the use of participatory tools in the communities. To what extent was the need for standardization of methodologies affecting participation levels and making the modules prescriptive rather than participatory? How to share information with participants and when to do it? How to deal with confidentiality and anonymity when reporting findings from participatory studies? And how to facilitate a process that empowered participants while acquiring information requested by national-level stakeholders? These issues are discussed in detail in Barahona and Levy (2003).

6.7 Concluding Thoughts

At the time that the SSC and Calibre Consultants were contracted to manage the Starter Pack M&E programme, many data sources in Malawi were discredited because of weak institutional arrangements that had affected standards of data collection and processing. There were serious problems of quality and reliability in relation to official data sources. The M&E programme made a major effort not only to ensure high standards at all stages of the process – from the design of the studies through training, field work and data processing to the presentation of information – but also to make our efforts transparent so that the strengths and weaknesses could be judged by end-users.

Rarely are the methods used by researchers described in enough detail to allow people outside the teams involved to make a fair assessment of the quality of the information provided. This leads to two sorts of problem: on the one hand, some decision makers use all data sources indiscriminately, without attention to their reliability; while on the other, some decision makers examine the data and find them to be unreliable, so they become cynical about all sources. Both reactions undermine the process of evidence-based decision making. The Starter Pack M&E programme attempted to address these problems.

However, Malawian institutional uptake of the lessons learnt by the M&E programme has been weak. There are some signs that the experience of the programme in terms of survey methodologies is being assimilated within the MoA where individuals who belonged to the M&E teams can use their expertise, and one team member has been promoted to head the Ministry's M&E unit. Some of the new participatory methods are being taught to students at the University of Malawi, where other M&E team members are based. However, the methodological innovations of the programme have not been fully assimilated by Malawian institutions. There are a number of reasons for this – including a perception in some quarters that the work done by the M&E programme is only of relevance to assessing the impact of Starter Pack/TIP. It is our hope that the overview provided in this chapter will enhance understanding of the wider relevance of our experience and innovations.

Notes

[1] The methodological innovations have been peer reviewed and two papers have been published: Barahona and Levy (2003) and Levy (2003). Since 2002, five workshops attended by a total of around 80 participants have been run to share these methodological experiences. Three workshops have been run at Reading University, one in Kenya and one in Uganda.

[2] The register compiled by the Logistics Unit included the names of all the intended beneficiaries and the villages where they live (see Chapter 3). This database provided an up-to-date list of villages throughout the country, which was considered a good sampling frame for villages. However, it was not reliable as a sampling frame for households within villages. For the surveys, we conducted complete household listings, while for the studies that used participatory methods, we developed an alternative approach to collecting numerical information that avoided the need for within-village sampling (see Section 6.5.4).

[3] This topic is developed in detail in Barahona and Levy (2003).

[4] Monthly price data for the main staple foods is available from FEWS-Malawi.

[5] A full description of the poverty index used in each year's evaluation surveys is provided in the relevant module reports (see CD in Back Cover Insert).

[6] A comparison group is similar to a control group. It is a group with similar characteristics to that of the group of interest – but it is not determined by the evaluation team.

[7] See Wingfield Digby (2000) for a discussion about definitions and classifications used by the Malawi census in 1998 and by those carrying out registration for Starter Pack.

References

Barahona, C. and Levy, S. (2003) How to generate statistics and influence policy using participatory methods in research: reflections on work in Malawi 1999–2002. Working Paper 212, Institute of Development Studies, Brighton, UK. (Available on CD in Back Cover Insert.)

Levy, S. (2003) Are we targeting the poor? Lessons from Malawi *PLA Notes* 47. International Institute for Environment and Development, London.

Levy, S. and Barahona, C. (2001) 2000/01 Targeted Inputs Programme (TIP): main report of the monitoring and evaluation programme. (Unpublished, available on CD in Back Cover Insert.)

Nyirongo, C.C., Gondwe, H.C.Y., Msiska, F.B.M., Mdyetseni, H.A.J. and Kamanga, F.M.C.E. (2002) 2001/02 TIP evaluation module 1: food production and security in rural Malawi (pre-harvest survey). (Unpublished, available on CD in Back Cover Insert.)

Nyirongo, C.C., Msiska, F.B.M., Mdyetseni, H.A.J., Kamanga, F.M.C.E. and Levy, S. (2003) Food production and security in rural Malawi (pre-harvest survey). Final report on evaluation module 1 of the 2002/03 Extended Targeted Inputs Programme (ETIP). (Unpublished, available on CD in Back Cover Insert.)

Sibale, P.K., Chirembo, A.M., Saka, A.R. and Lungu, V.O. (2001) 2000/01 Targeted Inputs Programme module 1. Food production and security. (Unpublished, available on CD in Back Cover Insert.)

SSC (2004) The monitoring and evaluation archive of Malawi's Starter Pack and Targeted Inputs Programmes 1999-2003. Statistical Services Centre, University of Reading, UK. (Available in Back Cover Insert.)

Wingfield Digby, P. (2000) Measuring the size of the rural population in Malawi. A contribution to the 1999/2000 Starter Pack evaluation programme. (Unpublished, available on CD in Back Cover Insert.)

7 Lessons on Management of Large-scale Research Programmes

SARAH LEVY

7.1 Introduction

This chapter looks at the Starter Pack/Targeted Inputs Programme (TIP) Monitoring and Evaluation (M&E) experience from a management perspective. Chapters 5 and 6 explained how the M&E programme was designed, how it was structured as a series of linked modules, how the research methods evolved and how the managers ensured that the evidence presented was of a high standard (technically reliable). This chapter examines in greater detail the management approach of the Statistical Services Centre (SSC) of The University of Reading and Calibre Consultants. The intention is to highlight a series of elements that are often neglected, but which need to be taken seriously when running large-scale data collection programmes to inform policy in developing countries.

7.2 Running Large-scale Research Exercises

A nationwide programme like Starter Pack requires a nationwide research exercise that can produce 'representative' results. Traditionally, this type of information is collected using survey methods. However, if certain statistical principles are observed, it is also possible for researchers to use participatory methods, which have advantages for collecting some kinds of information (see Chapter 6).

In both cases – surveys and research using participatory methods – the field teams must go to the selected sites, however difficult they may be to get to. Otherwise, the findings cannot claim to be nationally representative. The managers of the Starter Pack/TIP M&E exercise had a 'low tolerance' approach to requests for replacement sites: field teams were only allowed to replace a site if they could prove that it did not exist (a 'ghost' village) or was completely impossible to get to (e.g. it was cut off by flooding).

7.2.1 Standardization

From the manager's point of view, a key statistical principle that needs to be taken into account when running large-scale research exercises is that of standardization. In the case of surveys, there is well-established 'good practice' for survey design, fieldwork and data processing, which, if followed carefully, will produce standardized results. We found that particular care needed to be taken with:

1. *The translation of the questionnaire into local languages.* If the versions of the questionnaire in different languages – Chichewa and Tumbuka in the Malawi case – are not exactly the same, the information from different sites will not be comparable.
2. *Supervision of personnel.* In our early surveys in Malawi, we found serious problems with sub-standard work by both field staff and data entry clerks. We solved them by introducing strict control mechanisms: high consultant-to-supervisor and supervisor-to-enumerator ratios for field work; the inclusion of a 'consistency checking form' and a 'control panel' in each questionnaire (see Section 7.3.3); and a system of double data entry.

For research using participatory methods, the need for standardization means that in all sites: (i) research tools should be used in the same manner, and (ii) the same type of information (with the same level of detail) should be collected. The first condition implies the development of a field manual that specifies in detail the tools to be used and the steps to be followed for each activity throughout the period during which the research team interacts with the community in each site. The field manual is developed by the researchers in a process that involves discussion and appropriation of research objectives, preliminary testing and improvement of tools. It should be emphasized that in order to standardize the tools, a preliminary design phase is required. It is not possible to develop standardized Participatory Rural Appraisal (PRA) tools sitting in an office. Consultation with stakeholders at community level as well as testing of different approaches is essential for the development of the PRA tools.

The second condition is fulfilled by the use of a unique instrument for recording the information acquired in each site: a 'debriefing document'. This complements the field manual by providing a framework for recording the key information with the same level of detail across all sites.

7.2.2 Training

We also aimed to standardize the 'human element' as much as possible by careful training of field teams. Thorough training is essential when teams are to work in a large number of sites. The objective was that the teams should conduct in-village sampling and interviewing (in the case of surveys) or use the PRA tools and record the information (in the case of research using participatory methods) in the same way in all sites. In order to achieve this, all field staff (whatever their level in the hierarchy) needed to attend a single, centralized training session and to be present throughout. It is not recommended to have several teams trained in different places. This is because everybody should share the same interpretation and understanding of the approach. If there are variations from one field team to another, the results will not be comparable between sites.

7.3 Financial Planning and Management

Most research exercises are limited by the financial resources available. Often it is assumed by the proponents of surveys or of PRA that their method is inherently cheaper than the other approach. In our experience, such assumptions are unreliable: costs depend more on the nature of the study and on the value placed locally on the inputs required, in particular on human and other scarce resources. In Malawi, we found that the cost was similar for a nationwide survey using a simple questionnaire and for a participatory research study in 20–40 sites.

In the case of the Starter Pack/TIP M&E programme, the budget for each module (provided by the UK Department for International Development (DFID)) was £30,000–£40,000,[1] including local consultants' fees. It was part of the managers' job to get maximum benefit from these resources. This was achieved by a mixture of careful selection of research teams through competitive tendering, tight budgeting and financial controls, and a set of financial incentives to encourage timely, high-quality outputs. These issues are explored in the remainder of this section.

7.3.1 Competitive tendering

When we began working in Malawi, most of the research commissioned by donors was carried out by external consultants or by members of a small group of local consultants. There were high entry barriers for promising local consultants who were not part of the established group. In accordance with the capacity-building objectives of the M&E programme (see Chapter 5), we decided to open up the programme to all comers through a process of competitive tendering based on the 'Call for Proposals' approach used by DFID in the UK.

Thus, invitations to tender were issued for the first 2 years of the M&E programme. Consultants were invited to present a technical proposal and a budget for the research module for which they were bidding. The managers selected the best bid using an agreed – and public – list of criteria to evaluate the written proposals (including the budget) and to assess the short-listed teams at interview. This approach was found to have a number of advantages. On the financial side, it keeps costs down by encouraging bidders to be competitive in terms of their proposed use of resources. This was particularly important in relation to professional fees. In developing countries where foreign aid is a major source of finance in the economy, it is easy for donors to create a 'bubble' of inflated prices for professional services. By introducing competitive bidding, prices charged by consultants can be deflated to reasonable levels. Similar arguments apply on the technical side: competitive tendering encourages the best talent to come forward in an environment of equal opportunity, where selection is based on merit.

7.3.2 Budgeting and accounting

An important distinction in the module budgets was between consultants' fees and reimbursable expenditure. Consultants' fees were calculated on a daily rate basis. An estimate was made of the number of person days involved for each activity, but once the fees had been agreed, they were fixed – in effect they could be seen as a lump-sum payment for the work carried out. They did not vary unless the consultants failed to meet the terms of their contract by submitting sub-standard work or delivering a report after the agreed deadline (see Section 7.3.3). The reimbursable expenditure, on the other hand, was a forecast of the maximum expenditure allowable in each category. Actual expenditure was expected to be below this ceiling. Reimbursable expenditure was to be paid against receipt.

The managers found that consultants in Malawi had little experience with financial planning and budgeting for surveys. It was common to receive unrealistic budgets with inflated fees and allowances and a lack of attention to detail under reimbursable expenditure. It was necessary to insist that care be taken to produce realistic budgets.

The managers required accounts for reimbursable expenditure incurred to be presented at regular intervals throughout the project (see Section 7.3.4). The format for presenting the accounts was made as simple as possible, so that the accounts could be compiled by the consultants: the receipts under each category of reimbursable expenditure in the budget had to be collected, numbered sequentially and added up. The consultants were required to present summary sheets backed by numbered receipts; in the case of allowances, a list of names, amounts disbursed, dates received and signatures was to be attached. In effect, the accounts were simply a list of expenditures backed up by receipts or other documentary evidence. These could be subtracted from disbursements received to calculate the balance in hand.

> **Box 7.1.** The allowance problem.
>
> Many of the teams which won contracts under the invitation to tender for the Starter Pack/TIP M&E programme were associated with government institutions in Malawi. Therefore, when designing the system of payments for the M&E programme we had to be aware of the pay structures of government employees. Government employees with good professional qualifications were working for salaries of between US$60 and US$100 per month. With such low rates of pay for the jobs they were employed to do, they were forced to supplement their income by working for externally funded projects. Years ago, the donors introduced 'allowances' as a form of payment for such work. These were originally intended to cover subsistence costs when out in the field, but had become a form of disguised fee, since allowances had been inflated to well above costs. Moreover, allowances were provided for going into the field, but not for the product of the field visit – i.e. ensuring high standards of data collection and making sure that the data is processed, analysed and presented in the form of a report.
>
> The M&E programme broke with this pattern by making a clear distinction between fees and allowances for consultants.[2] Consultants' field allowances were established at a level which represented a real subsistence allowance, covering accommodation, meals and incidental expenses in the field. Their main payment was in the form of a fee.

7.3.3 Incentives

For this type of research exercise, it is essential to establish a set of clear, fair incentives, to explain them to all of those involved and to stick to them. All personnel need to be paid on time, with part of their payment withheld until after they have delivered a satisfactory product. Sub-standard and late work should be penalized and good work should be rewarded. A system which fails to create the right financial incentives is very unlikely to produce good quality information, even if the technical side of the study is faultless.

The M&E programme used a number of techniques to increase the chance of a successful outcome. For the managers' relationship with the consultants, these techniques included:

- Disbursement of funds in instalments, with amounts calculated to cover reimbursable expenditure *only* at the start of the project, and with the next instalment conditional on the consultants making progress on the technical side *and* on presenting satisfactory accounts.
- Disbursement of 80% of consultants' fees at the end of the project, on delivery of a satisfactory final report, with a 20% interim fee payment.
- Deductions from fees for late delivery of the final report or for any failure to ensure good quality data collection or processing, analysing and reporting of results. This provided a major incentive to deliver a good product, on time.
- Financial penalties for unauthorized off-budget expenditure (see Section 7.3.4), deducted from the consultants' interim or final fee payments.

The techniques used by the consultants with their teams were based on similar principles, although their implementation varied according to the financial management capacity of the consultants. By the time of the TIP surveys, each questionnaire contained a consistency checking form, which allowed the supervisors to check that the answers recorded by the enumerators were consistent, and the consultants (in their role as senior field supervisors) to do the same when collecting questionnaires from supervisors. The responsibilities of enumerators, supervisors, consultants and data entry clerks were reinforced by including in the questionnaires a 'control panel', which had to be signed by each person when their task was completed (see Table 7.1). Thus, enumerators would sign when they had completed a questionnaire; supervisors would sign when they had checked the work of the enumerator (using the consistency checking form); consultants would sign when they had checked the final product coming in from the field; and data entry clerks would sign after entering each questionnaire into the computer. Anybody could be financially penalized for failing to carry out their task properly – including the consultants, whose work was checked by the managers. Penalties generally took the form of having to redo the work without extra pay.

The distinction between allowances and fees for field supervisors and enumerators in surveys and for research assistants in PRA-type modules was more difficult to make than in the case of the consultants. The field staff continued to be paid the traditional type of 'allowances'. However, they were paid only 50% of their allowances up front; they received the other 50% when the work was completed. This suggested that half of the allowance was in fact a fee to be paid on delivery of a product. The final 50% (or part of it) could be withheld if the work carried out was not satisfactory.

7.3.4 Financial controls

The managers found that in Malawi, particularly in government departments, there was no culture of using resources efficiently. Some of our consultants appeared to be under the impression that donor funds were unlimited and required little accountability. The managers' role included close monitoring of expenditure by the teams. Accounts were demanded on a regular basis and reviewed carefully by the managers.

Table 7.1. Control panel.

	Name	Signature	Date
Enumerator			
Supervisor			
Consultant			
Data entry clerk			

Source: TIP evaluation surveys.

A key tool for financial control was the disbursement of funds in instalments. At any one time, we only provided enough funds to cover reimbursable expenditure for the period until the next presentation of accounts (1–2 months). Therefore, funds were low by the time the accounts were presented. Any problem in the accounts would imply a delay in receiving the next instalment of funds to allow the work to proceed. This provided a major incentive for consultants to present their accounts properly.

The managers also refused to pay for any unauthorized off-budget expenditure. They stressed from the beginning that all expenditure must be authorized by them in advance and in writing; otherwise it would be deducted from the consultants' fees. The same rules applied if the consultants wished to transfer funds from one budget line to another. At first these rules had little credibility, with consultants expecting the managers to be lenient and field staff expecting the consultants to be lenient. However, after one or two experiences of enforcement by the managers, the module teams began to take the rules seriously.

From the 2000/01 season, we also made it clear that the consultants' eligibility for future contracts would depend on their financial efficiency and probity. This provided a strong incentive, especially to the younger and more ambitious teams. One measure of success is that none of the 18 module teams spent more than the amount in their budgets.

7.4 Conclusion

The Starter Pack/TIP M&E programme aimed to provide policy makers with high-quality, timely information about a national-level intervention. This implied a large-scale research programme organized as a series of linked modules. At the same time, it involved local consultants, many of whom won contracts under the competitive tendering process despite having little previous experience of this type of M&E programme.

In order to manage such a programme successfully and ensure timely, high-quality research products, the SSC and Calibre imposed serious demands on the M&E teams in terms of visiting selected sites, following standardized procedures and financial management. Particular attention was paid to creating the right financial incentives and controls. These proved effective in supporting our goal of producing data and reports of internationally acceptable standards to tight deadlines within available budgets.

Notes

[1] At an exchange rate of US$1.6 to the pound, this was equivalent to US$48,000–US$64,000.
[2] An important issue – although one which was beyond our scope – is that the financial planning and management tools which improve the efficiency of resource management for large-scale research exercises should be applied within Malawi Government as well as by private consultants. This would imply introducing a proper salary and career structure for government employees and modern budgeting and accounting systems.

III Lessons from Starter Pack

8 Production, Prices and Food Security: How Starter Pack Works

SARAH LEVY

8.1 Introduction

Maize is the preferred staple food crop in Malawi, and is grown by almost all smallholder farmers.[1] In the 1970s and 1980s, the government encouraged smallholders to grow maize, but agricultural liberalization in the mid-1990s changed the economic context out of all recognition. Hybrid maize and fertilizer subsidies ended, and the government's credit scheme for smallholders all but disappeared. The exchange rate depreciated sharply after 1994, making imported inputs – particularly fertilizer – very expensive in local currency terms (see Introduction). This had serious consequences for maize production because smallholders had become heavily dependent on fertilizer and hybrid seed to boost yields.

This chapter draws on the evidence from the Starter Pack/Targeted Inputs Programme (TIP) Monitoring and Evaluation (M&E) programme to show that, in the post-liberalization context, Starter Pack makes a major contribution to national food security. However, the way it works is different from what the original designers of the programme expected. The chapter also considers broader questions about the role of Starter Pack in relation to agricultural growth, social protection and poverty reduction. Finally, it explores the issue of how long the programme will be needed and when 'exit' will be possible.

8.2 Maize Deficits and Food Insecurity in Malawi

Post-liberalization Malawi is a chronic under-producer of maize. According to the Ministry of Agriculture (MoA), smallholder production in the main season varied between 1.3 and 2.2 million t of maize in 1998–2003. In the absence of reliable population figures and estimates of production of other food crops[2] and winter-season maize, it is impossible to calculate national maize production deficits. However, it was clear from the food security outcomes that the 1999 and 2000 main-season harvests of over 2 million t of maize were large enough, and that the 2001 and 2002 harvests of under 1.5 million t were too small (Levy, 2003). This suggests that – in the absence of intervention – Malawi had a maize production deficit of some 500,000 to 600,000 t in the early 2000s.

The production deficit is apparent at household as well as national level. The TIP evaluation surveys show that few rural households are self-sufficient in maize from one harvest to the next. By 3 months before the harvest, which takes place in April–June, some three-quarters of households are without their own maize supplies even in a good year like 2000/01 (see Table 8.1). In 2001/02, the figure rose to 87%.

As rural households run out of maize, they may consume other home-grown foods. But other foods are not necessarily available when maize supplies run out. Nyirongo *et al.* (2003) found that the only major staples other than maize with production patterns allowing food stocks to be maintained throughout the year are cassava and bananas, but less than one-quarter of households have access to these food sources during the pre-harvest period.[3] Most try to access food off-farm at this time. The main off-farm sources of food are doing *ganyu* in return for maize, or purchasing food in the market.

However, when there are maize shortages, *ganyu*-for-food opportunities become scarce and it becomes difficult or impossible to buy maize and other foods in local markets (see Section 8.4). At such times, people resort to eating maize husks and cobs or foraging for wild roots, mushrooms and fruits. The strongly seasonal nature of food insecurity was documented by the TIP evaluation surveys using an indicator based on household coping strategies in times of hunger (see Chapter 6). It shows why the pre-harvest months are known in Malawi as the 'hungry period'.

8.3 The Starter Pack Contribution to Maize Production

The main reason for Malawi's chronic under-production of maize is poor soil fertility and a binding input constraint (see Chapters 9 and 10). The designers

Table 8.1. Months of maize deficit – smallholder farmers.

	2000/01 (% of farmers)	2001/02 (% of farmers)	2002/03* (% of farmers)
9 months or more	10	17	22
6 months or more	32	52	50
3 months or more	72	87	82
No deficit	5	3	6

*List A farmers only. Deficits were slightly higher for List B and non-beneficiaries.
Source: 2001, 2002 and 2003 TIP evaluation surveys.

of the Starter Pack programme argued that a small pack of fertilizer, high-yielding maize seed and legume seed would provide a large boost to maize production (see Chapter 1).

The Starter Pack/TIP evaluations were able to estimate the contribution of the programme to household maize production in the 1999/2000, 2000/01, 2001/02 and 2002/03 seasons (see Table 8.2). This was done by comparing volumes of maize produced by beneficiary households with volumes produced by a comparison group of non-beneficiaries[4] in the same geographical locations with similar poverty profiles.

Average household production of maize in a particular year depends on a number of external factors, in particular weather conditions. The weather was good in the 1999/2000 season and moderate in the following three seasons (parts of the country were affected by excess rainfall and flooding). However, weather conditions do not affect the contribution of the free inputs programmes as a percentage of household maize production.

Table 8.2 shows that Starter Pack is capable of contributing two to three 50-kg bags of maize per household. With the number of households in rural Malawi estimated at around 2.8 million (see Chapter 6), when one pack is distributed to each rural household – a 'universal' Starter Pack programme – the intervention is capable of producing between 280,000 and 420,000 t of additional maize at national level.

However, Table 8.2 also shows substantial year-to-year variations in percentage of average household production contributed by Starter Pack/TIP. In 1999/2000, Starter Pack contributed some 17% of beneficiary households' average maize output, but this fell to 9% in the 2000/01 TIP. The evaluation cannot fully explain these variations, but we can identify probable explanatory factors such as the timing of distributions and type of maize seed provided. The importance of timing of delivery is evident from the absolute figures for quantities produced in 2002/03, which show a very large difference between the outcomes for List A beneficiaries, who received their packs on time, and List B beneficiaries, who received them late. The other year when late delivery was a serious problem was 2000/01.

The issue of type of maize seed is more difficult to judge. There is much debate about whether hybrid maize seed is higher yielding in Malawi field conditions than open pollinated varieties (OPVs) (see Chapter 12). Beneficiaries of the 1999/2000 and 2002/03 (List A) free inputs distributions received hybrid seed, while in 2000/01 beneficiaries received OPV. At first sight, this would appear to substantiate claims that hybrid seed performs better than OPV. However, in the 2001/02 TIP, half the beneficiaries received hybrid seed, and there was no improvement compared with 2000/01.

8.4   Prices and Markets

The increased maize production from Starter Pack is not enough to make beneficiary households food self-sufficient. Smallholders in Malawi are often thought of as subsistence farmers, and policy makers talk about solving food

Table 8.2. Smallholder maize production and the Starter Pack/TIP contribution.

	Average household maize production from Starter Pack/TIP (kg)	Production from Starter Pack/TIP as percentage of beneficiaries' average household production	Number of packs of free inputs distributed (million)	Total maize production from free inputs ('000 t)	Total main-season smallholder maize production ('000 t)	Percentage contributed to total main-season maize production by Starter Pack/TIP
1999/2000	120	16.8	2.88	346	2212	15.6
2000/01	36	8.6	1.5	54	1495	3.6
2001/02	34	8.9	1.0	34	1319	2.6
2002/03 (A)	156	30.7	1.95	304	1626	18.7
2002/03 (B)	85	19.5	0.8	68		4.2

Notes: The table covers only the years in which evaluation surveys were run with the SSC's technical guidance and quality-control systems (see Chapter 6). For 2002/03, List A (first distribution) and List B (second distribution) are shown separately.
Source: Starter Pack/TIP evaluation surveys, except 'Total main-season smallholder maize production', which uses MoA crop estimate survey data from FEWS.

insecurity by promoting household food self-sufficiency. But few rural households are self-sufficient in post-liberalization Malawi; the majority are deficit food producers. Therefore, a key factor for food security is the price of maize. However great their need, households cannot access maize from off-farm sources (*ganyu* or local markets) during the hungry period if the price of maize is too high.

So what price of maize is affordable? What price is too high? It is not possible to give a precise answer to this question because of the lack of reliable data on rural incomes. However, we can use food security outcomes to get an idea about the impact of different maize price levels on rural consumers in the early 2000s:

> The food security outcomes in recent years suggest that prices of around MK5–10 per kg – as in the 1999/2000 and 2000/01 seasons – are affordable for most rural households. As the maize price rises above MK15 per kg, as it did in September–October 2001, observers in the field begin to report increases in food insecurity among the rural poor, and at MK30 or more, as in January–March 2002, there is likely to be a food crisis (Levy *et al.*, 2004).

Prices of maize (and other staple foods, which – as substitute goods – follow the trend in maize prices) are the key to understanding how Starter Pack contributes to enhanced food security in Malawi. But the mechanism is not straightforward. Volumes of maize supplied to local markets by smallholder producers (who account for roughly 90% of total maize output) are generally low (see Chapters 9 and 10), and variations in marketed volumes between good and bad harvest years are small. On the demand side, however, there can be sharp variations. In good years, most of the better-off households can feed themselves from on-farm sources (maize and other staple food crops). They can also provide *ganyu*-for-food opportunities for poorer households. However, in years that follow bad harvests, better-off households join the competition for the small amounts of maize available, prices rise sharply (see Figure 8.1) and poorer households are driven out of the market. In addition, wealthier households cannot provide enough *ganyu* for those who need it.

This is what happened in 2001/02, following the low maize harvest of 2001. The situation was worst in those areas where food sources are least diversified. Figure 8.2 shows that the strongest demand pressure was in districts that rely predominantly on maize. Districts with significant non-maize staple food crops (in the northern region, along the lakeshore and in the Shire River valley) experienced less demand pressure and lower food prices.

The key to the success of Starter Pack is that it reduces demand-side pressure in the market. Thus, food prices remain low throughout the hungry period. By contrast, strategies based on importing food work from the supply side. They can only achieve a similar impact in terms of fighting hunger if the imported maize is sold at affordable prices, implying – at the exchange rates that have prevailed in recent years – large domestic price subsidies (see Chapter 15). If maize imports had been sold at

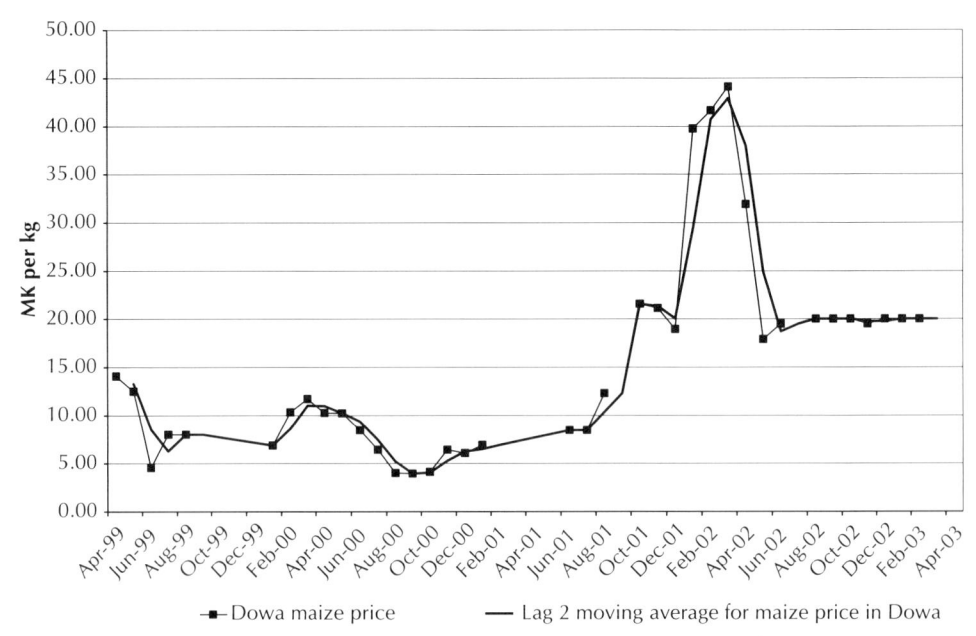

Fig. 8.1. Maize price in Dowa (central region), April 1999 to April 2003. (FEWS using MoA price data.)

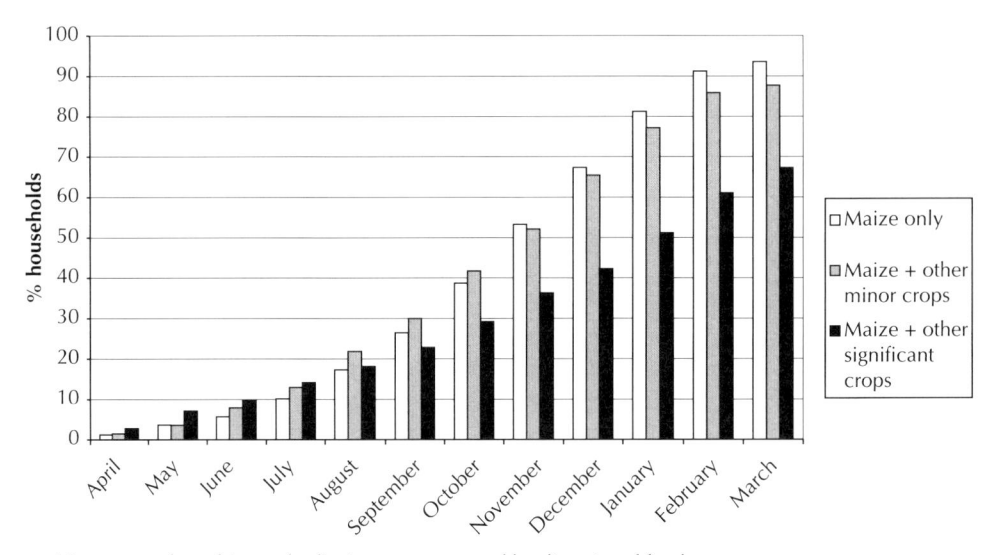

Note: The series is broken in places owing to field staff failing to collect/report the data.

Note: to produce this graph, districts were grouped by diversity of food crops grown.

Fig. 8.2. Proportion of households buying (or trying to buy) maize, 2001/02. (2001 and 2002 TIP surveys.)

import parity prices in early 2002, this would have implied a consumer price of MK20–30 per kg, which would have been unaffordable for most of the rural population.

8.5 The Importance of Scale

Because the food security impact of free inputs programmes operates through markets and prices, *geographical scale is important*. When Starter Pack is run on a large scale, reaching most smallholder farmers in most villages, even in the remotest parts of the country, it is an effective tool for achieving national food security. When it was scaled down to reach only one-half or one-third of smallholders in the 2000/01 and 2001/02 TIPs, the impact was disproportionately reduced. Although some households benefited from higher maize production (with 9% of their total maize output contributed by the free inputs), on aggregate the programme contributed only 3–4% of main-season smallholder production compared with 16% in 1999/2000 (see Table 8.2). This was not enough to dampen maize prices.

It is mainly for this reason that public works programmes (PWPs) – which provide cash, food or agricultural inputs as payment for work on public infrastructure – cannot achieve a similar national food security impact to Starter Pack. These programmes work well on a relatively small scale; in 2003, the largest PWP was the Malawi Social Action Fund (MASAF) with just over 98,000 beneficiary households. PWPs cannot be scaled up to reach most households in most villages on a regular basis, even if the resources are available to do so, as there are limited public works opportunities and management capacity.

8.6 Why 'Starter' Pack is a Misnomer in the Malawi Case

The designers of Starter Pack saw it as a programme capable of jump-starting growth in the agriculture sector (Chapter 1). They were optimistic that the programme would teach smallholder farmers a technology that would allow them to become surplus producers and make profits as commercial maize farmers. Free inputs would only be needed as a 'starter', after which farmers would be able to purchase them from commercial suppliers. This made sense given the prices of seed, fertilizer and maize that prevailed in the mid-1990s. It was possible to think in terms of a substantial proportion of the smallholder sector growing maize for sale. It was reasonable to think of providing farmers with free inputs for 2 or 3 years, demonstrating that they could increase yields so that they would be able to feed their families and produce a saleable surplus, and then phasing out the free inputs when they were in a position to buy them.

What the designers of the programme did not take into account was that this idea was vulnerable to changes in relative prices of inputs and outputs in the liberalized market environment. By 2003, increases in the price of fertilizer meant that virtually no smallholder farmer was able to make a profit from selling maize at prices compatible with food security in Malawi. Van Donge *et al.* (2001) reported that it was rare to hear a farmer reason in terms of price signals or profit margins, and the TIP surveys found only 10–15% of smallholder farmers selling any maize (see Chapters 9 and 10). In the present-day context, commercial maize farming is an unlikely strategy for smallholders, unless supported by large-scale, costly fertilizer price subsidies and credit schemes.

In the author's view, therefore, Starter Pack should not be seen as a growth strategy in the Malawi context. Nevertheless, the designers of the programme were right to point out that free inputs would help to achieve food security, which is essential for the success of growth, rural development and poverty reduction initiatives:

> Without securing the food supply, all other efforts at poverty alleviation – job creation, education reform, expanded health services – will come to naught. This is not a programme for recovering from drought. It is a programme to lay a solid foundation for long-term growth (Blackie *et al.*, 1998).

8.7 How Relevant are these Findings to Other Developing Countries?

It is worth pausing at this point to note the implications of the Malawi experience for other poor developing countries. We have argued that in Malawi, the rural smallholder population is characterized by large food-production deficits and vulnerability to price increases in the hungry period before the maize harvest. A tiny pack of free inputs can make a major difference to food security, if distributed on a large enough scale. The mechanism by which it achieves this is largely a demand-side one, and consists of dampening maize prices so that maize remains affordable for poor households throughout the hungry period. The food security contribution of Starter Pack is its greatest strength, and food security is an essential pre-condition for the success of any growth strategy.

The key element in this line of argument is the role of relative prices and farmers' reactions to them. In Malawi, we have argued that relative input–output price changes have made maize growing increasingly unprofitable, mainly owing to the increase in fertilizer prices in local currency over the last 10 years, mirroring the depreciation of the local currency against the dollar (see Introduction). This is beyond the control of the government (see Chapter 15).

In countries experiencing similar economic conditions in the post-liberalization context, this is a familiar story with similar implications. In such cases, 'Starter' Pack is a misnomer – but large-scale free inputs distributions could nevertheless be helpful in achieving national food security. On the other hand, for countries with stable macro-economic environments – in particular exchange rate stability – the original 'Starter' Pack concept may be a valid one. Nevertheless, even in such cases it is worth bearing in mind that a growth strategy that depends on producing a product (such as maize) that relies heavily on imported inputs (such as fertilizer) will be highly vulnerable to exchange rate shocks.

8.8 Social Protection and Poverty Reduction[5]

Malawi's free inputs programme has often been thought of as a safety net for the poorest and most vulnerable people in rural society, particularly after Starter Pack became the TIP. However, thinking on social protection is evolving rapidly, beyond the limited concepts of social welfare or assistance, and synergies with food security and agricultural growth are being identified (Devereux, 2003; Farrington *et al.*, 2004). This section suggests that attempts to make Starter Pack into a safety net failed, and ignored the real potential it offered for social protection.

8.8.1 Safety nets

During the early 2000s, the focus of the Malawi Government's social protection approach has been safety net interventions. The main components of its National Safety Net Strategy (NSNS) were conceived as welfare transfers, PWPs, child nutrition and free inputs for smallholder farmers. These were to be provided by the state to the poorest and most vulnerable members of society, particularly in rural areas. The 'target group' was estimated at 20–30% of the population (National Economic Council, 2000). A common theme of all components of the NSNS was to be poverty and vulnerability targeting, and this fed into the decision to scale down Starter Pack in 2000/01 and 2001/02 (see Chapter 2) and to try to use community targeting to select the poorest and most vulnerable farmers to receive packs.

8.8.2 Social protection synergies

Attempts to target free inputs to only the poorest and most vulnerable smallholder farmers – 'narrow' poverty targeting – failed (see Chapter 11), and they undermined the maize production and food security potential of the programme (see Section 8.5). Moreover, it soon became clear that the strongest contribution that free inputs programmes can make to social protection is via their wider food security role. A universal Starter Pack is an effective way of combating chronic food insecurity. The scaling down of Starter Pack contributed to the 2001/02 food crisis (see Introduction), which proved costly for the government and donors (see Chapter 15), but even more so for farmers. Many had to sell assets to survive, while some of the most vulnerable – such as the elderly and those already weakened by HIV/AIDS – did not survive the 2002 hungry season.

How does a universal Starter Pack contribute to social protection in the broad sense of reducing risk and vulnerability?

- It reduces the risk of food crises, which disproportionately affect the poor and vulnerable, providing a more stable foundation for such households to invest in agricultural growth.
- It strengthens traditional support systems within rural communities by increasing food availability (see Box 8.1).

> **Box 8.1.** Community support systems.
>
> People in rural Malawi are aware of a breakdown in traditional support systems. Some talk of the emergence of a 'table culture': whereas people would in the past eat communally under an open shelter, nowadays people eat inside their houses seated at a table. Shortage of food is the main reason for this decline. If there is not enough food to go around, people try to protect themselves by guarding what they have.
>
> One support system which is perceived to be on the increase is based on agricultural *ganyu*: the rich often create labour opportunities as a way of helping the poor, particularly those seeking work in exchange for food during the hungry period. But when food is short, the rich have fewer resources to pay for *ganyu*.
>
> Source: Adapted from Chinsinga *et al.* (2001).

Some additional advantages of the programme in comparison with other social protection interventions which have been implemented or piloted in Malawi are that:

- it provides support even in the remotest communities, which are often not reached by other initiatives;
- it is less costly than welfare transfer programmes (see Chapter 15) and has modest requirements in terms of management (see Chapter 3); and
- it is broadly poverty targeted and gender-neutral, unlike most PWPs, which (though not by design) tend to discourage participation by the poorest members of the community, particularly unaccompanied women.

Safety nets for the poorest and most vulnerable households (welfare transfers, PWPs and child nutrition programmes) will continue to be needed in Malawi, even with a universal Starter Pack programme. However, with Starter Pack in place, contributing to social protection, the number of people needing safety nets should be much smaller.

8.8.3 Poverty reduction

The Starter Pack/TIP research found that for rural communities in Malawi, the concept of poverty reduction was strongly associated with having enough food; and lack of food was seen as the hallmark of poverty (see Chapter 9). By enhancing food security, Starter Pack contributes to alleviating poverty, both directly – by allowing poor households to grow or access more food – and indirectly, by creating *ganyu* employment opportunities. However, it does not *reduce* poverty in the sense of having a lasting impact: few of the benefits of food security last beyond the agricultural year.

Nevertheless, food security is a necessary pre-condition for longer-term poverty reduction. Investments in agriculture, health, education and other areas of development are less effective if people are chronically malnourished. The 2001/02 food crisis was a timely reminder (for those who had forgotten the impact of drought in the early 1990s and the 1997/98 food crisis) that a food security shock is a major setback for government and donor-sponsored poverty-reduction efforts.

8.9 The Longer-term Perspective and the Question of 'Exit'

At present, rural Malawi is caught in a trap. Poverty is so extreme and widespread (see Chapter 16) and food security is so precarious that any shock is enough to cause a crisis. In order to break out of this trap, a two-pronged strategy is required: on the one hand, the country needs to produce enough food, and Starter Pack has proven to be an effective way of doing this; on the other hand, a longer-term growth, development and poverty-reduction strategy is needed. Only when reasonable progress has been made in both areas will it be possible to talk of 'exit' from Starter Pack in a positive sense (as opposed to 'getting out' of the programme). Levy and Barahona (2002) suggest that, as part of the longer-term approach, indicators should be agreed upon for measuring when an area should 'graduate' from receiving free inputs.

What type of longer-term growth, development and poverty-reduction strategy would be compatible with the demands of food security and the broader economic constraints? First, low prices of maize are critical for food security in rural Malawi, so a compatible growth, development and poverty-reduction strategy should not be based on high maize prices. Second, rapid exchange rate depreciation is likely to continue for the foreseeable future, so it is important to minimize dependence on imported inputs such as fertilizer. Third, the government's fiscal position is extremely precarious (Whitworth, 2004), so any strategy needs to take into account that government spending capacity is strictly limited.

For all of these reasons, a growth, development and poverty-reduction strategy based on commercial maize growing – which requires high maize prices, large volumes of imported fertilizer and heavy investment by government on subsidies, credit schemes and large-scale welfare transfers – is not a practical alternative at present. It would only become an option (albeit a risky one) should the government's fiscal position and the country's macroeconomic stability improve dramatically.

Although a discussion of the alternatives is beyond the scope of this chapter, it is worth highlighting the importance of:

- developing smallholder farmers' livelihoods by increasing opportunities for off-farm employment, business (trade) and crafts, promoting cash crops that can be grown without displacing food crops, and boosting livestock ownership;
- diversification of food and cash crops to reduce risk (see Chapter 12) and to minimize dependence on expensive imported inputs;

- development of input and output markets, which at present are uncompetitive and fail to reach farmers in the remoter rural areas; and
- improvement of rural infrastructure (especially roads, which provide rural communities with access to markets) and of security.

8.10 Conclusion

Starter Pack has proven to be an effective tool for combating chronic food insecurity and contributing to social protection in rural Malawi following agricultural liberalization. The degree of success achieved by the programme in this respect is remarkable: it is not often that we have evidence of such a positive outcome. It is to be hoped that policy makers in Malawi will take on board these findings and continue to implement the programme. However, Starter Pack – while a pre-condition for growth and poverty reduction in Malawi – is not sufficient to achieve these objectives. It needs to be accompanied by a longer-term, complementary strategy for development of rural livelihoods, food sources, markets and infrastructure. 'Exit' from Starter Pack will only be advisable when there are signs of progress towards growth, development and poverty reduction. Indicators should be designed to track progress towards these goals.

In post-liberalization Malawi, the way that Starter Pack works was found to be different from that anticipated by the designers of the programme. However, in other countries the original 'Starter' Pack concept may be more relevant. Anybody considering implementing a Starter Pack programme in another country should first evaluate the conditions prevailing there – including relative prices and exchange rate stability – in order to judge whether and in what manner Starter Pack could support food security, social protection, growth and poverty reduction.

Notes

[1] The other main staple food crops are cassava, sweet potatoes, rice, sorghum and bananas. The TIP evaluation surveys (for example Nyirongo et al., 2003) provide data on crop production.
[2] Estimates of production of cassava, sweet potatoes and bananas are a particular problem. No reliable production estimates have been produced in recent years.
[3] There are reports, however, that some households are now drying sweet potatoes and storing them for consumption during the pre-harvest period (see Chapter 14).
[4] In the TIPs, large numbers of smallholder households were deliberately excluded, but even when enough packs were distributed for every smallholder household in 1999/2000 and 2002/03, the surveys found substantial numbers of households that did not receive a pack.
[5] This section draws heavily on Levy et al. (2004).

References

Blackie, M., Benson, T., Conroy, A., Gilbert, R., Kanyama-Phiri, G., Kumwenda, J., Mann, C., Mughogho, S., Phiri, A. and Waddington, S. (1998) Malawi soil fertility issues and options – a discussion paper. MPTF/Ministry of Agriculture and Irrigation, Malawi. (Unpublished, available on CD in Back Cover Insert.)

Chinsinga, B., Dzimadzi, C., Chaweza, R., Kambewa, P., Kapondamgaga P. and Mgemezulu, O. (2001) 2000/01 TIP evaluation module 4: consultations with the poor on safety nets. (Unpublished, available on CD in Back Cover Insert.)

Devereux, S. (2003) Policy options for increasing the contribution of social protection to food security. ODI Forum Food Security in Southern Africa. Overseas Development Institute, London.

Farrington, J., Slater, R. and Holmes, R. (2004) Social protection and pro-poor agricultural growth: what scope for synergies? *Natural Resource Perspectives* No. 91. Overseas Development Institute, London.

Levy, S. (2003) Starter Packs and hunger crises: a briefing for policymakers on food security in Malawi. (Unpublished, available on CD in Back Cover Insert.)

Levy, S. and Barahona, C. (2002) 2001/02 TIP: main report of the evaluation programme. (Unpublished, available on CD in Back Cover Insert.)

Levy, S., with Barahona, C. and Chinsinga, B. (2004) Food security, social protection, growth and poverty reduction synergies: the Starter Pack programme in Malawi. *Natural Resource Perspectives* No. 95. Overseas Development Institute, London.

National Economic Council (2000) *Safety Net Strategy for Malawi*. Lilongwe, Malawi.

Nyirongo, C.C., Msiska, F.B.M., Mdyetseni, H.A.J. and Kamanga, F.M.C.E. with Levy, S. (2003) Food production and security in rural Malawi (pre-harvest survey). Final report on evaluation module 1 of the 2002/03 Extended Targeted Inputs Programme (ETIP). (Unpublished, available on CD in Back Cover Insert.)

van Donge, J.K., Chivwaile, M., Kasapila, W., Kapondamgaga, P., Mgemezulu, O., Sangore, N. and Thawani, E. (2001) 2000/01 TIP evaluation module 2.2. A qualitative study of markets and livelihood security in rural Malawi (Unpublished, available on CD in Back Cover Insert.)

Whitworth, A. (2004) Malawi's fiscal crisis: a donor perspective. DFID, Lilongwe, Malawi.

9 The Farmer's Perspective – Values, Incentives and Constraints

JAN KEES VAN DONGE

9.1 Introduction

The evaluation of the Starter Pack programme and the Targeted Inputs Programme (TIP) has given insights into the structure and dynamics of Malawian agriculture that are not available in other African contexts. Such insights are needed with increasing urgency as the links between economic liberalization, growth and poverty reduction become less and less self-evident. The following statement (United Republic of Tanzania, 2002) provides an example of how the challenge for policy makers is perceived:

> Changes in the level of poverty do not reflect the good macro-economic performance of the 1990s. This has raised questions about the macro–micro linkages in poverty reduction in Tanzania. The majority of the poor depend upon agriculture for their livelihood, and much more needs to be done to address the constraints that limit its growth and reduces its potential impact in reducing poverty.

This chapter gives insight into the micro–macro links in Malawian agriculture. It deals with the worldviews of smallholder farmers, which frame their decision making. Most of the evidence in the chapter is based on qualitative research at nine sites[1] (van Donge et al., 2001). The sites were purposively selected, but the findings are consistent with the evidence from the survey-based modules of the Starter Pack/TIP Monitoring and Evaluation (M&E) programme. The Malawi research poses interesting questions for other countries in southern Africa, where fewer studies of this kind have been carried out.

9.2 Attitudes to Growing Maize

There are major discrepancies between policy makers' discourse on the economics of maize production and food security, and the way these subjects are

©CAB International 2005. *Starter Packs: a Strategy to Fight Hunger in Developing Countries?* (ed. S. Levy)

perceived at the farm level. This section explores the dominant policy view in southern Africa about growing maize and its role in agricultural development and food security, and the implications of this view for perceptions of free inputs programmes. It then considers the evidence from the M&E programme about smallholders' attitudes to growing maize. The findings indicate a disjunction between the two worldviews, which may hinder effective policy making.

9.2.1 The policy makers' view

Increasingly, at the policy level, the answer to persistent problems of food insecurity is believed to lie in the emergence of a class of highly productive farmers who will grow maize for the market using high-level technology (Bingen *et al.*, 2003, Crawford *et al.*, 2003, Kelly *et al.*, 2003). In Malawi, this manifests itself in collaboration between the Sasakawa Global 2000 programme and Monsanto, promoting the use of fertilizer and hybrid seed in combination with the herbicide Roundup (Valencia and Nyirenda, 2003). A similar view underpinned much of the original thinking about Starter Pack (see Chapter 1). It argues that high-technology farming is profitable and that food security will result from it, provided there is proper market development.

In Malawi, this reasoning has led to efforts to encourage market development in rural areas through agri-dealer initiatives: partnerships between nongovernmental organizations (NGOs) (mainly US-based) and the Malawian private sector to set up rural shopkeepers as stockists of fertilizer and hybrid seed.[2] The main concern of such initiatives is the establishment of a commercially sustainable system of input provision. From this perspective, free inputs are, at best, a tool for teaching farmers to use the high-technology approach, so that they will move into commercial maize farming and buy inputs in future; at worst, they 'crowd out' private suppliers of agricultural inputs (see Chapter 10).

An additional concern on the part of policy makers is that free inputs promote 'dependency' of smallholders by creating a passive or lazy attitude of waiting for inputs to be provided rather than a proactive attitude towards obtaining them. A free gift will not be properly costed in the farm enterprise and therefore does not promote commercially sustainable practices.

9.2.2 The Malawian smallholder farmers' approach

These arguments refer to the mental maps of people: their feelings, thoughts and interpretations of the world. To find out whether policy makers' assumptions matched reality on the ground, van Donge *et al.* (2001) asked smallholder farmers their opinions about fifteen statements. The statements had contrary meanings in order to avoid leading to particular answers. For example: 'Not growing one's own food is a reason for shame' and '*ganyu* or business is a better way to get food than working on one's own land'. We asked respondents to indicate their degree of agreement with these statements and to comment on them. We elicited lively reactions in all sites. The responses showed a remarkably consistent and widely held set of sentiments.[3] In addition to the use of 'provocative statements', we also carried out in-depth interviews with farmers to document the nature of farm enterprises.

9.2.2.1 How do smallholder farmers see maize?
A striking element in the mental map of people in rural Malawi is the emotional value of maize. Eating maize is seen as essential to having a good life. Maize is the preferred food in most parts of Malawi, and when people talk of 'food' they are normally referring to maize. Only in Nkhata Bay did we find a disdain for maize and more appreciation for cassava.

Maize growing is seen as the heart of the farm enterprise, and receives special treatment. For example, one can see patches of yellow and stunted maize that are planted late and not fertilized. This is a waste of effort for poor households. However, when queried, the answer is 'everybody must show that one tries at least to grow one's own food'.

As a rule, maize is not grown with the idea of raising cash.[4] We found only one village, Kachuli (Dedza), where maize was grown for sale. Elsewhere, people only sold maize if they had more than enough to eat. If the supply of food was insecure in the household and it sold maize, this was seen as distress selling. This contrasted with attitudes to other staples, such as sweet potatoes, cassava and rice, which are grown partly, often primarily, for sale.

Pure cash crops like tobacco are often seen as a means to obtain money to grow maize. Orr (2000) found that the more farmers grew tobacco, the more they were food secure; if they had enough money from the previous year's tobacco farming, farmers would buy as much fertilizer and seed as possible to grow maize. People also talk openly about diverting inputs acquired on loan (for cash crop cultivation) to food crops. Engaging in business or crafts and doing *ganyu* are also seen as ways of making money to buy the inputs to grow maize.

It is also striking that maize is given a higher value than money. If they are hungry, poor people will sell cash crops (like tobacco) on extremely bad terms while they are still in the field in order to buy maize. Food attracts labour in the hungry season, and people often prefer to be paid in maize for *ganyu*, even when the cash equivalent of the payment in maize is less than the cash payment. We found that a rich household in the village is a household that has food, and this is also the basis for accumulation.

9.2.2.2 The importance of growing your own food
We wanted to find out about farmers' attitudes to growing their own food compared to receiving food as a free gift. Our starting point was that distribution of free inputs will only be worthwhile if farmers value the process of

growing their own food. The research found that pride in growing one's own food was a widely shared value, and it was usually seen as a reason for shame if a household did not grow its own food. For instance, we were told that:

> 'It is shameful when you do not have your own food because whenever you go around looking for maize to buy, people perceive you as a beggar who is totally desperate and stranded for food. This is unlike when you have your own food whenever you have need of it.' (Dedza)

A household that does not grow its own food is often associated with laziness and even theft:

> 'Without your own food production, you buy every food item, hence people start to point fingers at you. Some say that you are lazy, others say that you are rich.' (Dowa)
>
> 'You can be forced to steal from other people's gardens. You lose self respect as you are all the time admiring others who have food and comparing yourself negatively. There is no free maize for children to roast, hence they begin to beg and this is a very shameful thing.' (Karonga)
>
> 'Unless one is mad or lazy there is no reason for not growing one's own food. Such characters should not be tolerated in the village because they can easily turn into beggars and thieves.' (Mzimba)

Although many households in rural Malawi do not manage to grow enough of their own food (see Chapter 8), most respondents supported these sentiments. Thus, growing one's own maize is clearly important for social reasons.

9.2.2.3 Food, money and poverty

Lack of food was seen as the hallmark of poverty, which showed most clearly in responses to our statement 'People who do not grow their own food are not necessarily poor':

> 'This is not true because, in a village set up, most of the people that are poor are also those who do not grow their own food.' (Dedza)
>
> 'Poverty in villages is mostly associated with people who do not grow their own food or those that do not grow enough of the food crops.' (Mzimba)
>
> 'Richness lies in eating.' (Mulanje)
>
> 'Those who do not grow their own food are very poor, because they will have to buy everything including food.' (Nkhata Bay)

In contrast, money was seen as unreliable. For example, in Marko Mwenechilanga (Karonga) we were told:

> 'You may have money from business or *ganyu*, but the problem is that you may have nowhere to buy food. People in the village are reluctant to sell their food, especially maize.'

The responses to our statement 'Poor people are only really helped by money instead of free maize or more food on the farm' illustrate people's misgivings about money:

> 'Money makes a person restless. It can make someone go to South Africa, Blantyre and other places because at first it looks too much. By the time one will think of buying inputs, the money is gone.' (Nkhotakota)

> 'Money is so sweet (*inakometsa*) that one will end up buying almost each and everything. Hence the whole amount can finish within one day.' (Mulanje)

Cash has relatively little value for poor people in rural Malawi if they are hungry. They often do not have the means to travel to buy food on open markets and local access to food is often tied to work for larger farmers (see Section 9.3.3).

9.2.2.4 The question of laziness

At village level, free inputs were primarily perceived as activating the poor – encouraging them to grow more food – and not as leading to 'dependency'. Virtually nobody agreed with the statement that 'Free inputs make people lazy'. Vociferous condemnation was the usual response:

> 'It is only lazy people who think that Starter Pack makes people lazy. When people get the inputs, they make sure that they have put them into use. In fact, we add to the little that we have. We do not just wait for free inputs.' (Zomba)
>
> 'Actually, laziness comes when somebody does not have anything at all to apply to the crop.' (Mzimba)
>
> 'Free inputs make people work harder as their morale is boosted.' (Dedza)
>
> 'If people become used to free food or money, after the donation they will perish because in the process they will develop laziness. They have to work themselves to get their own food. Hence fertilizer and seeds are the best handout.' (Dowa)

The idea that free inputs create dependency assumes a natural tendency towards laziness among the poor. This was strongly denied in the responses. People generally felt that if the poor are inactive then it is because they do not have the resources to be active with. Indeed, staying alive is a struggle; you are likely to die if you are lazy.

9.3 The Smallholder's Production Constraints

The previous section shows that Malawi's smallholder farmers are convinced of the need for self-sufficiency, and welcome any intervention which allows them to work towards this goal. Failure to be self-sufficient means having to rely on purchasing food in the hungry period, when prices are high, or having to work for food. This is a risky situation to be in, and one that involves serious social stigma. Given this worldview, Malawian smallholder farmers' production decisions are about how to maximize food production and self-sufficiency. However, this goal is seldom achieved owing to the constraints faced by the majority of farmers, which are discussed in this section.

9.3.1 Agricultural decline and the land constraint

Van Donge *et al.* (2001) found that people perceived loss of soil fertility as the greatest problem facing agriculture in Malawi. Therefore, the fertilizer provided by the free inputs programmes was highly valued. This emerged particularly in reactions to our statements eliciting opinions about whether conditions have changed over the years: 'Life was better in the past because it was easier to get enough food from the land' and 'When I was a child, food was no problem as people were more interested in farming':

> 'Very true, because the soils were inherently rich in fertility and fertilizer was cheaper.' (Dedza)
>
> Life was better in the past 'because the land was fertile. These days people are also interested in farming, but you need fertilizer to have enough food.' (Mzimba)
>
> 'In the past, people lived better lives because they produced sufficient food. However, this time there is a lot of pressure on land due to overpopulation. Consequently, it has been eroded of its valuable fertility. People are still committed to farming, but they are let down by shortage of land and loss of soil fertility.' (Mulanje)
>
> 'Very true, land is limited and soil exhausted now, which means a low yield and a hard life.' (Dowa)

In individual conversations, soil fertility was mentioned as a major problem in all but two of the research sites. In Msilamoyo (Nkhata Bay) and Chisi (Nsanje) soil fertility is not yet a problem, but in focus group discussions people saw the problem looming.[5]

In some areas, production is also limited by the shortage of land available for cultivation. The evaluation surveys found that the average land area cultivated by smallholders in the main agricultural season is just over 2 acres, or approximately 1 ha (Nyirongo *et al.*, 2002, 2003). There are major regional differences, with land shortages at their most acute in the southern region. Levy (2003) reports that around three-quarters of List A[6] TIP beneficiary house-

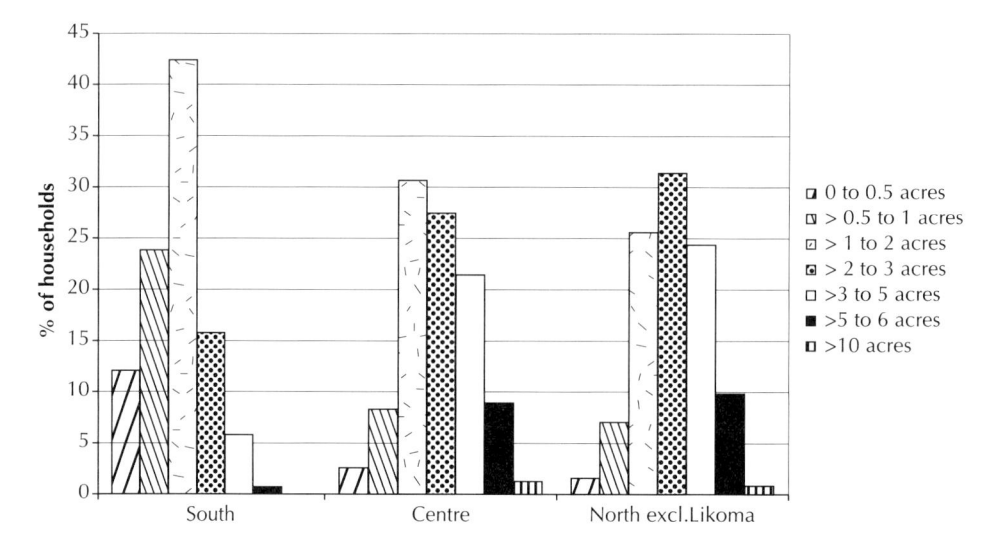

Fig. 9.1. Area cultivated by List A households, 2002/03. (Levy (2003) using data from the 2003 TIP evaluation survey.)

holds in the south cultivated 2 acres or less in the 2002/03 main season (see Fig. 9.1), compared with some 40% of households in the centre and only one-third in the north.

9.3.2 The capital constraint

Lack of capital to invest in crop production is a serious problem for most smallholder farmers. Income poverty – or lack of cash – means that most smallholders simply cannot afford to buy the fertilizer that they need to sustain maize yields in the context of declining soil fertility (see Chapter 10). The TIP evaluation surveys found a clear correlation between poverty/wealth and purchases of fertilizer (Nyirongo *et al.*, 2001, 2002, 2003).

9.3.3 Labour and the *ganyu* poverty trap

Malawi is a densely populated country, particularly in the south, suggesting that there should be an ample supply of labour. But this is not the case. There are large numbers of households comprising only elderly people and young children, or adults affected by sickness (an increasing proportion because of HIV/AIDS). In addition, labour demands in farming are concentrated at the time of planting and the weeding period. At these times, there is strong demand for labour on people's own farms; but if food has run out, household members have to look for work opportunities provided by wealthier farmers to stave off hunger. As a result, they plant late and neglect weeding, leading to much smaller harvests. Thus, although labour tends not to be mentioned as a constraint by farmers themselves, it is clearly a factor that limits the production capacity of the poor.

The most common form of work opportunity in rural areas is the type of agricultural labour referred to as *ganyu* in Chichewa;[7] it may be defined as:

> *An agreement to do a spatially delimited piece of farm work for an agreed compensation. The essential thing is that there is no payment unless the work is completed and the employees may to a large extent organize their own time.*

Doing *ganyu* is clearly associated with poverty. This is because:

> 'Food from *ganyu* is very unreliable, as the payment is very small. They pay you only a small plate and yet you have a family to feed. For *ganyu* you cultivate a very big acre and they pay you food that lasts only for a short period, unlike if you cultivate the same area yourself, then you can harvest a lot.' (Karonga)
>
> '*Ganyu* is a tough work associated with little payment and much suffering. Not many people would voluntarily choose to do *ganyu*. One cannot depend upon *ganyu* for the rest of his or her life to get food, let alone for the family. *Ganyu* labour is not always available as it has specific seasons, notably lean months. Hence, what will happen in the off-season?' (Dedza)

Thus, *ganyu* was seen as an indication that someone is trapped in extreme poverty. It is associated with a loss of control over your own livelihood. As a focus group in Mzimba said: 'Depending on *ganyu* shows lack of direction'.

Box 9.1. *Ganyu* and other work for food.

Mrs Kaombe of Mkalo village (Machinga) is a widow whose two children have left home. She grows her own maize, sorghum and cassava, but normally she runs out of food by December. When she has no more food, she survives by winnowing rice and maize for other people at the mill. The remains, especially broken rice, are payment in such a case. It is quite humiliating to do this work. Competition is bitter, and the people who bring their crops for milling act as bosses. 'The hungry do literally everything for them.' She used not to go for *ganyu*, but in the last growing season she had to do so in order to get food. She got a small pail of maize for 2 weeks' work. 'It was a kind of punishment and a lesson at the same time.'

Source: Adapted from van Donge *et al.* (2001).

9.4 Cash Crops vs. Maize Growing

The previous sections have presented evidence to show that smallholder farmers in Malawi do not grow maize for profit, but to avoid the risk of hunger and social stigma. However, in order to grow maize, they must battle with a number of production constraints. While it is difficult to do much about land and labour constraints, the problem of low maize yields can be dealt with if the farmer can overcome the capital constraint, buying fertilizer and improved seed.

Thus, the logical strategy for the smallholder farmer is to try to increase income. Everywhere in rural Malawi, households are involved in a drive to generate cash. The sale of crops is the most important source of cash in all poverty categories, so agriculture plays a central role in income generation (see Chapter 10). However, maize is relatively unimportant in crop sales (see Fig. 9.2). Farmers rely more on 'cash crops' such as tobacco, cotton, vegetables and rice, and on crops which serve both to generate income and as food sources, such as sweet potatoes, cassava and legumes. Cash crops are often concentrated in specific geographical areas, as in the case of tobacco. Tobacco farmers are an elite who earn high incomes from crop sales.

Cultivation of cash crops is often seen as being in the service of maize production, and the amount of cash generated determines the amount of inputs that can be used to produce food. Thus, the effective demand for inputs is for a large part motivated by food security considerations. The paradox is that people need to be involved with the market to buy inputs in order to feel free of the market through food self-sufficiency.

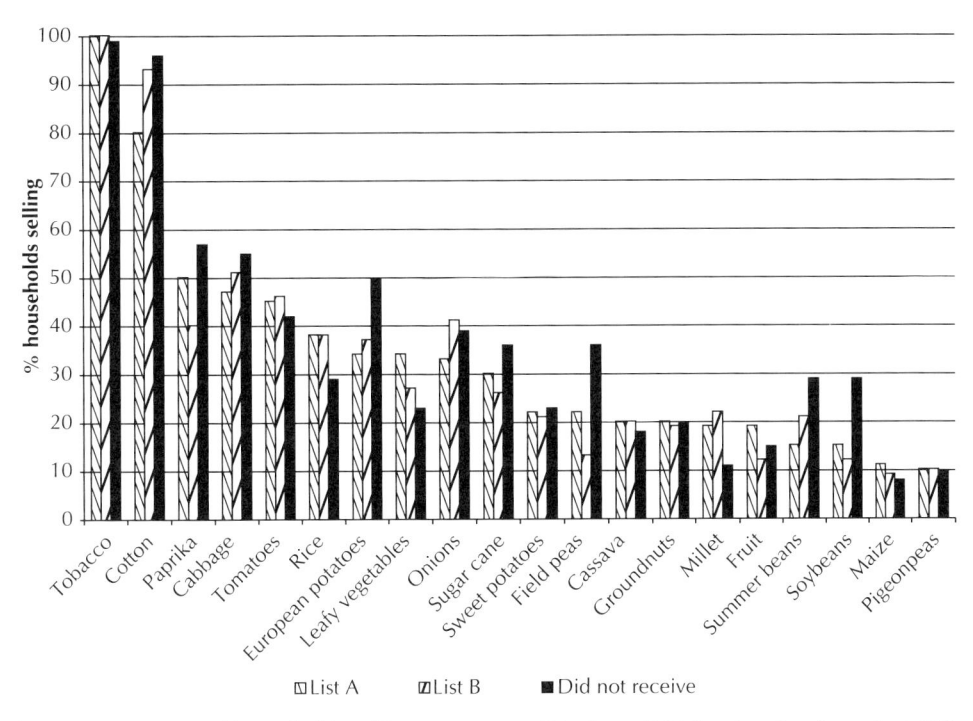

Fig. 9.2. Proportion of households selling crops, 2002/03. (Levy (2003) using data from the 2003 TIP evaluation survey.)

9.5 Conclusion

The Starter Pack programme appeals to strongly held values in rural Malawi. A good and respectable life in the village means growing one's own food. Lack of food is the most excruciating aspect of poverty. Distribution of free inputs is seen as an adequate way to alleviate poverty and to restore self-respect.

The smallholder farmers' approach is often closely related to a distrust of money and the perception that home-grown food is a protection against the treacherous sphere of monetary relations. Yet, at the same time, the involvement with the world of money is seen as unavoidable. Household self-sufficiency in food may be of great value, but this is impossible without inputs: fertilizer and hybrid seed. In order to buy inputs, households need to generate income, and the main income-generation strategy is cultivation of cash crops.

The assessment of smallholder agriculture has changed considerably in influential policy discourse over the years. Whereas peasants were seen as a major historical force in the 1960s, they are now considered marginal in the world economic order (Bryceson, 2000). Smallholder agriculture that values subsistence is – in the evolutionary view of input supply – seen as regressive

(Crawford *et al.*, 2003). From here it is only a small step to dismiss smallholder farming as inefficient production. The World Bank's report (2000) 'Can Africa Claim the 21st Century' declared that African peasant agriculture is not competitive internationally. However, unless there is sufficient growth in other sectors to absorb the labour force in smallholder agriculture, many rural households will continue to depend upon their smallholdings for survival. The distribution of free inputs may not transform smallholder agriculture, but this chapter has argued that it fits well with the needs and aspirations of smallholder farmers.

Notes

[1] The villages were: (1) Marko Mwenechilanga in Karonga; (2) Msilamoyo village in Nkhata Bay; (3) Tombolombo village in Mzimba; (4) Kamange village in Nkhotakota; (5) Katsakunya village in Dowa; (6) Kankodola village in Dedza; (7) Mkalo village in Machinga; (8) Thopina village in Mulanje and (9) Chisi village in Nsanje. The research was carried out by three teams consisting of a graduate and a university student: Prince Kapondamgaga and Elarton Thawani; Noel Sangore and William Kasapila; Overtoun Mgemezulu and Mackenzie Chivwaile. The teams spent around 2 weeks in each site.

[2] Three examples are: (1) the Rural Market Development initiative of the Citizens Network for Foreign Affairs (CNFA), funded by The Rockefeller Foundation; private-sector participants are the seed and fertilizer companies. (2) The Agricultural Input Markets Development Project (AIMS), funded by USAID. (3) The Sustaining Productive Livelihoods through Inputs For Assets (SPLIFA) project launched in the 2003/04 season by the IFDC, CARE International (Malawi) and other NGOs with funding from DFID.

[3] In this chapter we only produce majority opinions. These majorities were usually overwhelming, although there were some dissenting views, which gave important insights.

[4] This is in line with the theory of the 'normal surplus' as developed by Allan (1965). Allan tried to explain erratic output patterns of maize and the lack of relationship between output and price among the plateau Tonga in southern Zambia.

[5] Orr *et al.* (2000) also note the importance of soil fertility. They report on a pest control management project in Thyolo which found that soil fertility – not pest management, as assumed – was the fundamental constraint in agriculture.

[6] Distribution of TIP inputs was done in two phases in 2002/03; the figures are for TIP beneficiaries who received the inputs in the first phase (List A beneficiaries). The findings were similar for second-phase TIP beneficiaries (List B) and non-beneficiaries.

[7] Whiteside (2000) wrote a seminal paper on *ganyu*. He used a much broader definition than the one used here. For example, he includes working on estates. This makes sense if the work involves seasonal piecework and is not wage labour or share cropping. More problematic is his inclusion of work parties – those where everyone takes turns working on each other's farms as well as those where people are working for cooked food or beer – as in *chipere ganyu*. We have not heard these practices referred to as *ganyu* and we found that wherever such cooperative arrangements were strong, there was less *ganyu* labour (Marko Mwenechilanga, Karonga; Msilamoyo, Nkhata Bay).

References

Allan, W. (1965) *The African Husbandman.* Oliver and Boyd, Edinburgh, UK.

Bingen, J., Serrano, A. and Howard, J. (2003) Linking farmers to markets: different approaches to human capital development. *Food Policy* 28, 405–419.

Bryceson, D. (2000) Peasant theories and smallholder policies. In: Bryceson, D., Kay, C. and Mooij, J. (eds) *Disappearing Peasantries? Rural Labour in Africa, Asia and Latin America.* Intermediate Technology Publications, London.

Crawford, E., Kelly, V., Jayne, T.S. and Howard, J. (2003) Input use and market development sub-Saharan Africa: an overview. *Food Policy* 28, 277–292.

Kelly, V., Adesina, A.A. and Gordon, A. (2003) Expanding access to agricultural inputs in Africa: a review of recent market development experience. *Food Policy* 28, 379–404.

Levy, S. (2003) Starter Packs and hunger crises: a briefing for policymakers on food security in Malawi. (Unpublished, available on CD in Back Cover Insert.)

Nyirongo, C.C., Gondwe, H.C.Y., Msiska, F.B.M., Mdyetseni, H.A.J. and Kamanga, F.M.C.E. (2001) 2000/01 TIP evaluation module 2.1. A quantitative study of markets and livelihood security in rural Malawi. (Unpublished, available on CD in Back Cover Insert.)

Nyirongo, C.C., Gondwe, H.C.Y., Msiska, F.B.M., Mdyetseni, H.A.J. and Kamanga, F.M.C.E. (2002) 2001/02 TIP evaluation module 1: food production and security in rural Malawi (pre-harvest survey). (Unpublished, available on CD in Back Cover Insert.)

Nyirongo, C.C., Msiska, F.B.M., Mdyetseni, H.A.J. and Kamanga, F.M.C.E. with Levy, S. (2003) Food production and security in rural Malawi (pre-harvest survey). Final report on evaluation module 1 of the 2002/03 Extended Targeted Inputs Programme (ETTP). (Unpublished, available on CD in Back Cover Insert.)

Orr, A. (2000) 'Green Gold'? Burley tobacco, smallholder agriculture, and poverty alleviation in Malawi. *World Development* 28(5), 347–363.

Orr, A., Mwale, B., Ritchie, J.M., Lawson-McDowall, J. and Chanika, S.S.M. (2000) *Learning and Livelihoods: the Experience of the FSIPM Project in Southern Malawi.* Natural Resources Institute of the University of Greenwich, Greenwich, UK.

United Republic of Tanzania (2002) Research and analysis working group. *Poverty and Human Development Report 2002.* Mkuki na Nyota Publishers, Dar es Salaam.

van Donge, J.K., Chivwaile, M., Kasapila, W., Kapondamgaga, P., Mgemezulu, O., Sangore, N. and Thawani, E. (2001) 2000/01 TIP evaluation module 2.2. A qualitative study of markets and livelihood security in rural Malawi. (Unpublished, available on CD in Back Cover Insert.)

Valencia, J.A. and Nyirenda, N. (2003) The impact of conservation agriculture technology on conventional weeding and its direct effect on maize cost production in Malawi. Sasakawa Global 2000, Lilongwe, Malawi.

Whiteside, M. (2000) Ganyu labour in Malawi and its implications for livelihood security interventions – an analysis of recent literature and implications for poverty alleviation. *Agricultural Research and Extension Network* 99. Available at: www.odi.org.uk/agren/publist1.html

World Bank (2000) *Can Africa Claim the 21st Century?* The World Bank, Washington, DC.

10 Do Free Inputs Crowd Out the Private Sector in Agricultural Input Markets?

CLEMENT NYIRONGO

10.1 Introduction

It is clear that free inputs programmes make a positive contribution to maize production and household food security in Malawi. However, there is concern about the programmes' impact on the agricultural input markets. Do free inputs prevent the emergence of a sustainable private trade in seed and fertilizer? This chapter examines the arguments and the empirical evidence on this subject collected by the Targeted Inputs Programme (TIP) Monitoring and Evaluation (M&E) teams in the 2000/01, 2001/02 and 2002/03 seasons.

The key concern expressed about free inputs programmes is that they 'crowd out' the private sector. The concept of crowding out in this context implies that something which would have been bought from the private sector – seed and/or fertilizer – is no longer purchased because of the intervention (i.e. because the inputs have been provided free of charge by the government). This hinders the development of the private sector.

How can we find out if inputs which would have been bought from the private sector are no longer being bought because of the free inputs programme? Our approach was to compare beneficiaries and non-beneficiaries with similar poverty profiles, to find out if the beneficiaries were buying less fertilizer and seed than the non-beneficiaries. The evidence presented in this chapter shows that this was not the case for fertilizer. For seed, there was a small degree of crowding out of private suppliers.

It is important to note that the crowding out argument makes two key assumptions: (i) that private supplies of inputs are available; and (ii) that demand for inputs would exist in the absence of the intervention.

However, these assumptions are only partially valid in the case of rural Malawi. The evidence shows that the real problems facing the development of a private trade in fertilizer and seed are related to supply-side market imperfections and weak demand.

10.2 Is There Any Evidence of Crowding Out?

In this section, we compare data on purchasing of fertilizer and maize seed by TIP beneficiaries and non-beneficiaries. A direct comparison can be made because the poverty profiles of the two groups are almost identical (see Chapter 11). The aim is to establish whether beneficiaries bought less fertilizer and seed than non-beneficiaries, which would indicate that 'crowding out' occurred as a result of the intervention.

10.2.1 Fertilizer

The 2002/03 main-season evaluation survey collected data on purchases of fertilizer by smallholder farmers. Nyirongo *et al.* (2003) found no evidence of crowding out:

> . . . around 30% bought fertilizer in 2002/03 and there are no significant differences in the proportions of TIP beneficiaries and non-beneficiaries buying. The similarity between the two groups implies that free fertilizer distribution did not deter TIP recipient households from purchasing fertilizer. This was because the pack provided only a tiny amount of fertilizer (10 kg), which is not enough even for the small cultivated areas of smallholder farmers. Clearly, those who could afford to buy fertilizer were not put off doing so by the receipt of a TIP pack.

Nyirongo *et al.* (2003) also found that volumes of fertilizer bought were similar for TIP beneficiaries and non-beneficiaries – or slightly higher for TIP beneficiaries (see Table 10.1). The mean amounts spent on fertilizer by non-beneficiaries were slightly higher than the mean amounts spent by beneficiaries, but the median amounts spent are almost the same, indicating that the patterns of expenditure were similar except among the high spenders.

In the previous 2 years, there were found to be very small differences between recipients and non-recipients:

- In 2000/01, 26% of TIP recipients bought fertilizer compared with 32% of non-recipients (Nyirongo *et al.*, 2001).
- In 2001/02, 23% of TIP recipients and 28% of non-recipients bought fertilizer with cash, and the median amounts spent were similar (Levy and Barahona, 2002).

10.2.2 Maize seed

The 2002/03 main-season evaluation survey also collected data on purchases of maize seed by smallholder farmers. Table 10.2 shows that a slightly lower

Table 10.1. Average amount of fertilizer purchased by smallholder farmers, 2002/03.

	Region	Group 1: those buying > 10 kg		Group 2: those buying ≤ 10 kg	
		Mean (kg)	No. of farmers	Mean (kg)	No. of farmers
TIP beneficiaries	North	112.3	82	8.2	5
	Centre	181.3	158	7.7	15
	South	80.8	168	6.8	30
	Malawi	126.0	408	7.2	50
Non-beneficiaries	North	126.5	39	8.9	7
	Centre	140.6	80	7.7	14
	South	78.0	112	7.3	10
	Malawi	107.9	231	7.8	31

Note: Only farmers who bought fertilizer were included. Distribution of TIP inputs was done in two phases; the figures for TIP beneficiaries are for those who received the inputs in the first phase (List A beneficiaries).
Source: Nyirongo et al. (2003).

proportion of TIP beneficiaries purchased maize seed than non-beneficiaries, but the differences are small.

Table 10.3 shows that in 2002/03, TIP beneficiaries who bought maize seed also spent less, on average, than non-beneficiaries. The differences between mean and median expenditure are not large, although they are slightly larger for improved seed than for local seed.

The findings of 2002/03 confirm the pattern of the previous 2 years: in 2000/01, 15% of beneficiaries and 24% of non-beneficiaries bought improved seed, while in 2001/02, the figures were 12% and 22%, respectively (Levy and Barahona, 2002). Thus, the evidence suggests that TIP did depress demand for maize seed, particularly improved varieties.

We conclude that TIP does have some crowding out impact on maize seed suppliers, but not on suppliers of fertilizer. In the following sections, we attempt to place the crowding out argument in perspective by examining the

Table 10.2. Use and purchase of maize seed, 2002/03.

	% households			
	TIP beneficiaries		Non-beneficiaries	
Type of maize seed	Used	Purchased	Used	Purchased
Local maize seed	81.8	26.4	81.7	29.2
Improved maize seed	97.7	8.3	25.6	11.9
Total respondents	1524		859	

Note: Distribution of TIP inputs was done in two phases; the figures for TIP beneficiaries are for those who received the inputs in the first phase (List A beneficiaries).
Source: Nyirongo et al. (2003).

Table 10.3. Smallholder expenditure on seed, 2002/03.

	\multicolumn{6}{c}{Median (MK)}					
	TIP beneficiaries			Non-beneficiaries		
Type of maize seed	Mean (MK)	Median (MK)	No. of responses	Mean (MK)	Median (MK)	No. of responses
Local	196	150	358	206	188	226
Improved	566	400	129	687	450	107

Note: Distribution of TIP inputs was done in two phases; the figures for TIP beneficiaries are for those who received the inputs in the first phase (List A beneficiaries).
Source: Nyirongo et al. (2003).

much greater challenges of supply-side market imperfections and weak smallholder demand for farm inputs.

10.3 Market Imperfections

The evidence from the M&E programme shows that rural households in Malawi in the first decade of the 21st century face a serious input constraint. The surveys of the 2000/01, 2001/02 and 2002/03 main agricultural seasons consistently found that *less than one-third of smallholders buy any fertilizer, and less than one-quarter buy improved maize seed*. This is not because farmers are unaware of the benefits of using these inputs. It is because of a combination of problems, which result in situations where either the inputs are not available (often the case with maize seed), or they are available but in larger volumes and at higher prices than farmers can afford (the case of fertilizer). Therefore, the key policy questions are: How can we make sure that sufficient supplies of inputs are available to smallholders? And what can be done to enable smallholders to purchase them?

10.3.1 Agricultural liberalization and reform

In order to understand the current position of smallholder farmers, it is necessary to understand how conditions have changed in the agriculture sector in recent years. Before 1985, the state intervened in the marketing of agricultural inputs for smallholders through the Agricultural Development and Marketing Corporation (ADMARC), a state-owned agricultural marketing agency established in 1971 by a Parliamentary Act. Government imported farm inputs and ADMARC sold them to farmers on a cash or credit basis at subsidized prices. In 1988, the government established the Smallholder Agricultural Credit Administration (SACA) as a department of the Ministry of Agriculture (MoA) to provide smallholder farmers with credit to buy farm inputs through ADMARC. The 1980s saw an increase in fertilizer use, facilitated by improved access to farm inputs, credit and price subsidies. However, even during the period of input subsidies and easy credit, only some 30% of smallholder farm households used fertilizer.

In the mid-1980s, it was realized that the government-controlled input supply system was neither efficient nor sustainable. The government agreed to agricultural liberalization, which included gradual removal of input price subsidies to create a level playing field for private traders. Fertilizer and maize seed subsidies were phased out completely in 1994/95.

With the change in political climate in the early 1990s, smallholder agricultural credit was politicized, repayments fell and the credit system collapsed in 1993 because of high default rates. In 1995, the Malawi Rural Finance Company (MRFC) was set up to re-establish credit discipline and improve loan recovery. Nevertheless, the rural credit system remains under considerable strain due to under-capitalization, high interest rates and high levels of delinquency.

10.3.2 Supply-side constraints

Malawian smallholders have had difficulty adjusting to the changed conditions. The expected benefits of liberalization have failed to materialize because of infrastructure bottlenecks, uncompetitive markets and adverse macro-economic conditions. The World Bank (2003) reported that fertilizer prices in Malawi are nearly three times the world prices and at least 20–50% higher than those in neighbouring landlocked countries. It found that the cost of international transport and port charges through Beira was about US$75/t; local transport added US$25/t; finance charges for 6 months were about 20% of total cost; trading company margins ranged between 25% and 40%, due to lack of competition and high exchange rate risk; and retailer margins were around 10%.

In 2003, according to data from the IFDC, fertilizer was sold to farmers at a retail price of US$16–17 per 50-kg bag. Rapid depreciation of the Malawi Kwacha added to the problems of high transport and financing costs, risk premia and business margins, with the local currency retail price of basal and urea increasing 17-fold and 20-fold respectively between 1994 and 2003 (see Introduction).

The TIP evaluation surveys found that availability of fertilizer was not a problem in rural areas. The main problem was high price relative to farmers' cash availability or purchasing power (see Section 10.4). A related problem was the minimum amount of fertilizer available for sale. Fertilizer is normally sold in 50-kg bags. Despite a recommendation by the team which originally designed Starter Pack that small bags of fertilizer (1–3 kg) should be made available in all rural markets (see Chapter 1), it appears that this had not happened – at least not on a large scale – by the 2002/03 main season. Table 10.1 shows that few farmers bought fertilizer in quantities of 10 kg or less.

For seed, on the other hand, there is a problem with availability (see Table 10.6). The difficulty of obtaining seed implies that the improved seed market is relatively underdeveloped compared with the fertilizer market. This is surprising when one considers that all inorganic fertilizers are imported while most improved seed is produced in Malawi. There are currently three key players in the seed industry, all private companies: Seed Co, Monsanto and Pannar. It is possible that they do not find it worthwhile to supply some of the remoter rural markets, as relatively few smallholders buy improved seed (see Table 10.2). However, if availability is perceived as a problem by farmers, demand is clearly outstripping supply. Policy makers could tackle this problem by facilitating increased availability of seed through non-market mechanisms if private suppliers are unable to meet demand.

10.4 Demand-side Problems

Smallholder farmers are aware of their need for fertilizer and seed. Van Donge *et al.* (2001) found that in most of the country 'the decline in soil fertility appears to be the major constraint on farming'. One farmer in Mzimba observed that in the past 'the land was fertile. These days people are also interested in farming, but you need fertilizer to have enough food' (Van Donge *et al.*, 2001). Cromwell *et al.* (2000) ranked indicators of agricultural sustainability in 30 villages throughout Malawi, and found that seed availability was perceived by farmers as one of the three most important indicators.[1]

However, a high level of *need* for fertilizer and seed – however well understood by farmers – does not translate into a high level of *demand*. The main reason for this is poverty. The results of the Integrated Household Survey (IHS) of 1998 showed that 60.6%[2] of the country's rural population live below the poverty line (National Economic Council, 2000).

There is clear evidence that the poverty status of a household affects its capacity to purchase fertilizer. Table 10.4 shows that wealthier households bought much more fertilizer than poor households in 2002/03. A strong correlation between poverty/wealth and fertilizer use was also found in previous years' TIP evaluation surveys. It can be concluded that the poverty status of the household has a significant impact on the level of demand for farm inputs. The key policy implication is that promoting farmer income levels will improve effective demand for fertilizer (see Section 10.5).

The problem of being unable to afford farm inputs comes up again and again in the responses to the surveys. Table 10.5 shows that lack of cash and the price of fertilizer were the main reasons why two-thirds of smallholder households did not buy any fertilizer in 2002/03. Availability was not a major obstacle. For seed (all crops), availability was mentioned as a problem slightly more often than price in 2002/03[3] (see Table 10.6).

However, in the case of both seed and fertilizer, smallholder farmers' demand will remain relatively weak while lack of cash (income poverty) is their main problem. This problem is discussed further in the next section. A related issue is the lack of access to credit since the collapse of the country's

Agricultural Input Markets

Table 10.4. Quantity of fertilizer purchased in 2002/03, by poverty.

Poverty category	TIP beneficiaries Mean quantity (kg)	No. of farmers	Non-beneficiaries Mean quantity (kg)	No. of farmers
1: Poorest	44.8	13	57.6	9
2: Poorer	53.5	36	28.7	27
3: Poor	47.9	70	80.5	54
4: Less poor	73.8	139	80.8	80
5: Least poor	178.3	200	141.9	92

Note: Only farmers who bought fertilizer are included. Distribution of TIP inputs was done in two phases; the figures for TIP beneficiaries are for those who received the inputs in the first phase (List A beneficiaries). For an explanation of the poverty categories used here, see Chapter 6 (Section 6.5.2).
Source: Nyirongo *et al.* (2003).

credit system. Nyirongo *et al.* (2001) found that fewer than 2% of smallholders had access to credit for buying fertilizer, while less than 1% had access to credit for purchasing improved seed. With limited cash and even more limited credit, the majority of smallholder farmers clearly cannot afford to buy inputs.

10.5 What Hope for the Future?

We have suggested that crowding out by TIP is not as important an obstacle to the expansion of the market for seed and fertilizer as supply-side market imperfections and weak demand. In this section, we first examine whether, despite the problems identified, there is any evidence that private businesses are expanding their share of the (limited) market. We then consider what measures might be implemented to overcome the problems and foster the development of the agricultural input markets.

Table 10.5. Reasons for not purchasing fertilizer, 2002/03.

	TIP beneficiaries	Non-beneficiaries
Price of fertilizer	44.2	36.5
Availability of fertilizer	8.1	8.0
Change in income from maize sales	0.1	0.2
Change in income from other crop sales	2.5	2.0
Change in income from other sources	3.9	2.5
Change in crops grown	1.2	3.2
Availability of cash	83.3	80.0
Soils do not need fertilizer	8.0	10.6
Total responses (those who did not buy)	1007	559

Note: Distribution of TIP inputs was done in two phases; the figures for TIP beneficiaries are for those who received the inputs in the first phase (List A beneficiaries).
Source: Nyirongo *et al.* (2003).

Table 10.6. Reasons for not purchasing seed (all crops), 2002/03.

	TIP beneficiaries	Non-beneficiaries
Price of seed	29.5	28.7
Availability of seed	38.7	35.3
Change in income from maize sales	0.2	0.5
Change in income from other crop sales	1.8	1.1
Change in income from other sources	3.9	3.5
Change in crops grown	4.4	5.5
Availability of cash	62.7	68.6
Total responses (those who did not buy)	1001	558

Note: Distribution of TIP inputs was done in two phases; the figures for TIP beneficiaries are for those who received the inputs in the first phase (List A beneficiaries).
Source: Nyirongo et al. (2003).

10.5.1 Private traders' share of the market

The proportions of TIP beneficiaries mentioning ADMARC as a source of fertilizer has declined in recent years (see Table 10.7). The figures for non-beneficiaries are similar. An increasing proportion of smallholders are sourcing their fertilizer from private suppliers – mainly private traders, but also other farmers. Thus, the evidence shows that the private sector is gradually increasing its share of the total fertilizer market. A similar trend can be observed for smallholder purchases of improved seed between the 2000/01 and 2001/02 seasons. However, in 2002/03, there was a decline in purchases from private sources, probably because the 2002 food crisis had undermined private traders' capacity to supply, owing to diversion of maize seed to be sold as food earlier in the year.

10.5.2 Overcoming supply-side constraints

If private sector involvement in agricultural inputs marketing is to be promoted, the following limitations need to be addressed:

- The key problem for fertilizer is cost. Policy makers should consider establishing associations of producers, importers, distributors and retailers. These associations would share information, organize bulk purchases and coordinate to cut costs.
- As international transport costs constitute a large part of the final cost of fertilizer, serious consideration should be given to the development of the Nacala corridor, which offers the shortest route to the sea.
- Macro-economic policy should take into account the needs of smallholder farmers, in particular with regard to the exchange rate, as rapid depreciation of the Kwacha tends to undermine all other efforts to reduce the cost of fertilizer.

Table 10.7. Sources of inputs for smallholder farmers (TIP beneficiaries).

	% farmers		
	2000/01	2001/02	2002/03
Source of fertilizer			
Purchased from ADMARC	16.2	11.3	7.7
Purchased from private traders	5.8	9.4	18.9
Purchased from other farmers	1.2	2.4	3.5
Source of improved seed			
Purchased from ADMARC	7.0	5.4	1.9
Purchased from private traders	5.6	6.2	4.8
Purchased from other farmers	1.7	2.1	1.3

Note: Distribution of TIP inputs in 2002/03 was done in two phases; the figures are for those who received the inputs in the first phase (List A beneficiaries).
Source: Nyirongo *et al.* (2001, 2002, 2003) and database of the 2001/02 evaluation.

- With regard to seed, the problem of availability needs to be addressed. It may be necessary to increase the number of companies that produce improved seed. Another way of increasing improved seed supply would be to encourage large-scale farmers to grow and market open pollinated varieties (OPVs).
- Efforts should also be made to encourage marketing of inputs in small bags, which are affordable for cash-poor smallholders.

10.5.3 Boosting demand

The most difficult problem of all for policy makers is probably that of weak demand, because overcoming this problem entails reducing poverty, particularly income poverty. The first step is to examine smallholders' income sources, so that a coherent livelihood strategy can be designed to develop these sources. Improvements in farmers' income will lead to increased purchases of farm inputs.

Figure 10.1 shows that smallholder farmers' income sources are diverse. The main sources are sale of crops, *ganyu* wages and small business. Results from the M&E programme show that sale of crops constitute a primary source of income, but the role of maize in generating income is small and declining. In 2002/03, only some 10% of smallholder farmers sold any maize, compared with 14% in 2001/02 and around 20% in 2000/01. Most farmers grow maize for food, not for sale (see Chapter 9). Maize has a low market ratio (proportion of the crop produced that is sold), indicating that it is not a cash crop.

What policy implication can be drawn from this? If farmers' purchasing power is to be strengthened, it is important to promote their capacity to grow cash crops, such as tobacco, cotton, vegetables and rice, and food-and-cash

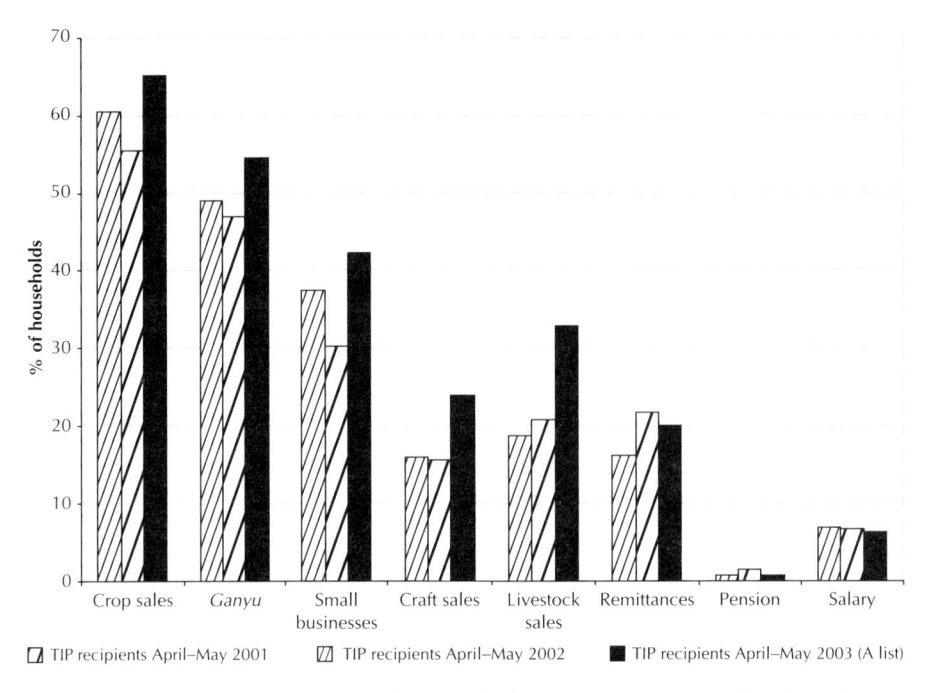

Fig. 10.1. Sources of income for smallholder households. (Levy (2003) using data from the 2001, 2002 and 2003 TIP evaluation surveys.)

crops, such as sweet potatoes, cassava and legumes, and to keep livestock (ownership of farm animals is extremely low in Malawi). Efforts should be made to support development of these income sources, e.g. through improved access to markets and better prices for producers.

The other main income sources – *ganyu* and small business – also deserve attention. However, *ganyu* wages (as opposed to wages from employment or salaries) represent a survival mechanism. People seek *ganyu* when they have no decent source of livelihood (see Chapter 9). If they were able to earn decent incomes from other sources, *ganyu* would decline as an income source. Nevertheless, there is a distinction to be made between types of *ganyu*, as some types of non-agricultural *ganyu* which do not compete with the planting or weeding seasons represent more attractive sources of livelihood.

Small businesses are also important income sources, and are often linked to agriculture. It is very common to see small businesses booming in rural areas during the harvest and post-harvest period and then closing down during the next growing season.

Finally, two further issues also need to be addressed:

1. Consumer price inflation should be kept under control. Farm inputs (an investment good) compete with consumer items in the basket of goods and services that the farm household has to spend money on. Rapid inflation, by diverting limited cash resources into consumption, affects demand for farm inputs.

2. Land reform issues need the attention of policy makers, as demand for inputs is partly dependent on land area available for cultivation. Farmers will not purchase more inputs than they can use on the land to which they have access.

10.6 Conclusion

Despite understandable concerns about crowding out by free inputs programmes, the evidence presented here shows that crowding out explains little if any of the difficulties facing private traders of agricultural inputs in Malawi. Private sector involvement in inputs markets is principally constrained by other factors. These include the income poverty of the majority of smallholders relative to the price of fertilizer and of basic consumption goods and services, and poor availability of seed in the remoter rural areas.

Weak demand resulting from low incomes and the high price of fertilizer is probably the most important constraint facing private sector expansion in the agricultural input markets. Structural problems relating to the supply of inputs are also hindering market development. Nevertheless, private traders are gradually gaining ground from the parastatal ADMARC.

Rather than focusing their energies on criticizing free inputs programmes, those who wish to promote the development of private input markets should tackle the problems of supply-side constraints and weak demand. Once these problems are solved, and smallholder farmers are able to access the inputs they need without the support of government, it should be possible to phase out the free inputs programmes.

Notes

[1] Crop diversification ranked first at national level, farmland size ranked third, and fertilizer application sixth (Cromwell et al., 2000).
[2] This comes from Table 5 (National Economic Council, 2000), which presents data from the 'more reliable' dataset containing information from a sub-set of 6586 households.
[3] In the previous year's survey (Nyirongo et al., 2002), price of seed was mentioned more often (34% of households) than availability (25% of households). It is probable that seed was particularly scarce in the 2002/03 planting season, following the food crisis. However, both factors are clearly important in both years: 2001/02 and 2002/03.

References

Cromwell, E., Kambewa, P., Mwanza, R. and Chirwa, R. with KWERA Development Centre (2000) 1999/2000 Starter Pack evaluation module 4: the impact of Starter Pack on sustainable agriculture in Malawi. (Unpublished, available on CD in Back Cover Insert.)

Levy, S. (2003) Starter Packs and hunger crises: a briefing for policymakers on food security in Malawi. (Unpublished, available on CD in Back Cover Insert.)

Levy, S. and Barahona, C. (2002) 2001/02 TIP: main report of the evaluation programme. (Unpublished, available on CD in Back Cover Insert.)

National Economic Council (2000) *Profile of Poverty in Malawi, 1998: Poverty Analysis of the Malawi Integrated Household Survey, 1997/98*. National Economic Council, Lilongwe, Malawi.

Nyirongo, C.C., Gondwe, H.C.Y., Msiska, F.B.M., Mdyetseni, H.A.J. and Kamanga, F.M.C.E. (2001) 2000/01 TIP evaluation module 2.1. A quantitative study of markets and livelihood security in rural Malawi. (Unpublished, available on CD in Back Cover Insert.)

Nyirongo, C.C., Gondwe, H.C.Y., Msiska, F.B.M., Mdyetseni, H.A.J. and Kamanga, F.M.C.E. (2002) 2001/02 TIP evaluation module 1: food production and security in rural Malawi (pre-harvest survey). (Unpublished, available on CD in Back Cover Insert.)

Nyirongo, C.C., Msiska, F.B.M., Mdyetseni, H.A.J. and Kamanga, F.M.C.E. with Levy, S. (2003) Food production and security in rural Malawi (pre-harvest survey). Final report on evaluation module 1 of the 2002/03 Extended Targeted Inputs Programme (ETIP). (Unpublished, available on CD in Back Cover Insert.)

van Donge, J.K., Chivwaile, M., Kasapila, W., Kapondamgaga, P., Mgemezulu, O., Sangore, N. and Thawani, E. (2001) 2000/01 TIP evaluation module 2.2. A qualitative study of markets and livelihood security in rural Malawi. (Unpublished, available on CD in Back Cover Insert.)

World Bank (2003) Malawi: country economic memorandum – policies for accelerating growth, January 2003.

11 Practical and Policy Dilemmas of Targeting Free Inputs

BLESSINGS CHINSINGA

11.1 Introduction

After 2 years of universal Starter Pack, the programme was scaled down in 2000/01 and 2001/02, becoming the Targeted Inputs Programme (TIP) (see Introduction; Chapter 2). The original Starter Pack concept of *broad* poverty targeting, where the target group was all smallholder farmers in Malawi, was replaced by a *narrow* poverty targeting concept under the TIP: only the 'poorest of the poor' – seen as the most deserving households – were to receive packs.

If beneficiaries were to be the poorest smallholder households, and wealthier smallholders were to be excluded, the programme had to devise a way of selecting the poorest households. The approach adopted was a form of community targeting, which aims to empower communities to identify beneficiaries themselves. The idea is that communities are better able to identify the poor than are outsiders, particularly government authorities.

As part of the Monitoring and Evaluation (M&E) programme, a series of surveys and participatory studies were carried out to assess the extent to which community targeting was successful in identifying the poorest beneficiaries. This chapter presents some of the lessons learnt about poverty targeting in free inputs programmes. These are important for those considering implementing such programmes in future in Malawi or other developing countries. The evidence shows that community targeting was not successful. The main lesson emerging from the Malawi experiment is that community targeting is complex and difficult to do, especially in circumstances where poverty data is either crude or non-existent. It is achievable with the help of external facilitation, but the human and financial resource demands of such facilitation are too great for large-scale input distribution programmes.

11.2 The Policy and Theoretical Context of Targeting

In recent years, targeting has become a key issue in the design and execution of interventions dealing with poverty, food insecurity and vulnerability (Besley and Kanbur, 1993; Gwatkin, 2000). It is a concern for many developing countries, especially those undergoing macro-economic stabilization and structural adjustment programmes, where governments are pressured to reduce expenditure.

Efficiency and effectiveness are an important part of targeting. Interventions should be capable of reaching those most in need. From the targeting perspective, an ideal intervention is one that not only accurately identifies the poor but also directs the benefits to them without any leakage (Harrigan, 2001; Barrett and Clay, 2003).

There is much debate about how benefits can be delivered to the most deserving people in order to achieve the greatest reduction in poverty, food insecurity and vulnerability. Several targeting strategies are distinguished in the literature (Hoddinott, 1999; Gwatkin, 2000). The main ones are: administrative targeting, self-targeting, geographical targeting and community targeting.

Administrative targeting involves government or non-governmental organization (NGO) employees in selecting beneficiaries. In the first 2 years of Starter Pack, Ministry of Agriculture (MoA) extension staff were asked to register beneficiaries, but this approach was quickly abandoned. Even though the packs were only *broadly* targeted, the extension workers felt that their involvement made them unpopular and undermined their extension role (see Chapter 3). A self-targeting approach involves making the benefit available to anyone who is prepared to provide labour (e.g. on a public works programme). It is has been considered as an alternative for Starter Pack/TIP, but the idea has been abandoned because it is impossible to provide work opportunities in more than 30,000 villages throughout Malawi. Geographical targeting has greater attractions from the technical point of view, but faces political obstacles (see Section 11.7.3).

The final option – community targeting – appears to be the most attractive alternative for targeting free inputs, and was the approach adopted by the TIP. The concept is briefly reviewed in the next section, in order to set the context for the rest of the discussion.

11.2.1 Community targeting

Community targeting gives communities the responsibility of deciding on the beneficiaries of an intervention, often through some kind of a representative structure such as a village committee or task force. It has become popular in recent years following the advent of democracy in many developing countries. According to Nelson and Wright (1995), it derives from the notion of participation as an end, not just as a means. The lead role of communities in targeting exercises is viewed as part of the development process, as it helps to build local decision making capacity. It gives communities the opportunity to wrestle with tough decisions about the allocation of scarce resources to alternative uses (Adato and Haddad, 2002).

The rationale for community targeting is that local people know best who is poor and which households are suffering a crisis or have the least income earning potential. Community involvement is seen as a way to 'exploit specialized local knowledge for targeting' – knowledge that is not available to outsiders (Ravallion and Wodon, 1998).

11.2.2   Challenges in targeting

The fully efficient targeting scenario is one in which it is possible to identify accurately all the deserving beneficiaries for an intervention and weed out the rest with equal precision. However, the targeting outcome seldom achieves this. 'Errors of inclusion' and 'errors of exclusion' are common (Hoddinott, 1999; Gwatkin, 2000). Errors of inclusion occur when an intervention reaches individuals who were not intended to be beneficiaries. Errors of exclusion arise when intended beneficiaries are missed out by an intervention.

Besley and Kanbur (1993) point out that successful poverty targeting 'presupposes agreement on what is meant by poverty – agreement on (i) a measure of the standard of living, (ii) a poverty line that distinguishes the poor from the non-poor, and (iii) a poverty index that aggregates information on the standard of living of the poor'. Even if the other conditions are met, data deficiencies are likely to give rise to errors in targeting. In most cases, the poverty data is not readily available and, if it exists, it is too crude to be useful for targeting purposes.

11.3   Community Targeting Mechanisms in the TIP

How was community targeting organized for the TIP? Village task forces (VTFs) were entrusted with the responsibility of selecting the beneficiaries to receive packs (Lawson *et al.*, 2001; Chinsinga *et al.*, 2002). The VTFs, it was emphasized, had to be as representative as possible of the interests within villages. They were to include members of all the main political parties, traditional and religious leaders and prominent persons of good standing. The village head was usually a key member of the VTF.

The VTFs were the implementing agency within the institutional hierarchy set up for beneficiary selection. At the summit of the hierarchy were the District Task Forces (DTFs), which united a number of Area Task Forces (ATFs). The DTFs coordinated the process, while the brief of the ATFs was to facilitate the formation of the VTFs and to inform communities either directly or through the VTFs about the process of beneficiary selection and the underlying spirit and rationale for community targeting. The ATFs outlined to the VTFs the guidelines on how to select the most deserving beneficiaries. For the 2001/02 TIP, for instance, the criteria for selecting beneficiaries were:

- widows/widowers with no source of income;
- the aged without any support in terms of financial resources but capable of using the pack; and
- families keeping orphans without support.

Those who did not satisfy the criteria were to be excluded.[1] It was expected that strict adherence to these selection criteria would ensure the efficiency and effectiveness of the process and lead to success in targeting the poor.

11.4 Design of the Evaluation Studies

The main challenge for the evaluation studies was to find out whether community targeting had selected the most deserving households as TIP beneficiaries. To achieve this, they had to compare the poverty levels of beneficiary and non-beneficiary households.

The evaluation surveys profiled the poverty status of beneficiary and non-beneficiary households using a poverty index derived from a combination of assets and income (see Chapter 6). Households were placed in five categories on a continuum stretching from the very poor to the least poor. The 'poverty profile' of beneficiary households could be compared with that of non-beneficiary households (see Section 11.5).

The evaluation studies based on participatory methods[2] also compared the poverty status of beneficiary and non-beneficiary households. This was done using a technique based on social mapping, which the evaluation managers called 'community mapping with cards' (see Chapter 6). In the Chinsinga *et al.* (2002) study, participants were asked to make a card for each household on the map and to place the cards in three categories according to the food security status of the household which the card represented: food secure, food insecure or extremely food insecure. Thus, food security was used as a proxy measure for poverty (Levy, 2003). The study used predefined categories to ensure comparability of the results across sites. The participants were also asked to identify which households were TIP beneficiaries and which were not. The information about food security and beneficiary status was recorded on each household card. The data on the cards were aggregated by the researchers and comparisons were made between beneficiary and non-beneficiary households (see Section 11.5). An innovative feature of these participatory studies was that they were able to generate statistics for use at a much more generalized level than is normally the case.[3]

11.5 Results of the Evaluation Studies

The survey data comparing the poverty status of beneficiary and non-beneficiary households (see Tables 11.1 and 11.2) indicate that the TIP failed to target the poor effectively. Targeting would have been successful if almost all of the beneficiaries and none of the non-beneficiaries were in poverty cate-

Table 11.1. Poverty profiles in 2000/01.

Poverty category	Beneficiaries (%)	Non-beneficiaries (%)
1:Poorest	29.9	23.7
2:Poorer	16.4	14.9
3:Poor	16.1	13.9
4:Less poor	13.5	14.9
5:Least poor	24.0	32.6
Total	100.0	100.0

Source: Adapted from Lawson *et al.* (2001).

gories 1 and 2. However, in both 2000/01 and 2001/02, beneficiaries were spread across all poverty categories – poorest to least poor – in much the same way as non-beneficiaries.

Some of the quantitative results of the participatory studies are presented in Table 11.3. If poverty targeting had been successful, there should not have been any food secure TIP beneficiaries or any extremely food insecure non-beneficiaries. However, the Chinsinga *et al.* (2002) study found that in 2001/02 there was only a slight preference for extremely food insecure households among the beneficiaries; many beneficiaries were food secure (21%), while 27% of non-beneficiary households should have received TIP, as they were extremely food insecure.

The implication of these results is that the poverty targeting process was marred by considerable inclusion and exclusion errors. A good number of the non-poor were included as beneficiaries and many of the deserving poor were excluded.

11.6 Why Did Poverty Targeting Fail in the TIP?

This section examines why community poverty targeting failed in the TIP. It draws on the findings of the evaluation studies based on participatory research methods.

Table 11.2. Poverty profiles in 2001/02.

Poverty category	Beneficiaries (%)	Non-beneficiaries (%)
1:Poorest	22.8	19.7
2:Poorer	21.2	20.5
3:Poor	21.6	22.2
4:Less poor	16.7	19.3
5:Least poor	17.7	18.3
Total	100.0	100.0

Source: Adapted from Nyirongo *et al.* (2002).

Table 11.3. Correlation between receipt of TIP and food security status in 2001/02.

Food security status	Beneficiaries (%)	Non-beneficiaries (%)	Total (%)
Food secure	21.2	33.5	28.9
Food insecure	38.5	39.7	39.3
Extremely food insecure	40.3	26.8	31.8
Total	100.0	100.0	100.0

Source: Chinsinga *et al.* (2002).

11.6.1 Resistance to targeting and social tension

Two of the participatory studies (Chinsinga *et al.*, 2001, 2002) simulated beneficiary selection in order to find out whether participants were able and willing to select the poorest households.[4] The studies found strong resistance to the idea of targeting, which was seen as an alien concept. Communities were reluctant to exclude some households from qualifying for the TIP, arguing that 'we are all poor and we all need assistance'. People in rural Malawi see themselves as enjoying the same status and therefore believe that they are equally entitled to external assistance. The notion of targeting invoked fears of precipitating discord in the social fabric of the communities and of witchcraft (see Box 11.1).

The reluctance of communities to distinguish themselves into poverty categories has been observed elsewhere in Africa. Hoddinott (1999) reports on a similar experience in Mali. A household survey indicated that there was a degree of economic differentiation between households in one village, but this was strongly denied during subsequent participatory group discussions. Hoddinott observes that communities often resist targeting, as it 'might be seen as creating or exacerbating social tensions within villages'.

Box 11.1. Resistance to targeting at Kachilili village, Mchinji district.

When the notion of targeting was introduced, the participants immediately pointed out that it could not be entertained in their community. They said that those who were left out would not take it kindly because they are all poor. Moreover, the selection of some households, leaving out others, might lead to fights and even cases of witchcraft.

When further urged by the researchers to undertake the beneficiary selection exercise, they singled out six of the 77 households in the village as households that could, in their view, be excluded. They said the rest should qualify as beneficiaries. They concluded by asking the research team to report that 'all the people of Kachilili village are poor so they cannot do the selection exercise'.

Source: Adapted from Chinsinga *et al.* (2001).

11.6.2 The spirit of egalitarianism

The reluctance to distinguish the poor in rural communities in Malawi may be understood in the context of the communities' egalitarian values, particularly with regard to food production (Chinsinga *et al.*, 2002, 2004). The evaluation teams found that communities quickly adapted the targeting process to fit their own perceptions and interpretations of equity, need and entitlement. A variety of redistributive measures were encountered. The case presented in Box 11.2 exemplifies the egalitarianism of rural Malawi.

This situation is not unique to Malawi. Adato and Haddad (2002) report on the forces of egalitarianism in South Africa in the context of a public works programme. Here, communities were entrusted with the responsibility of hiring people to work on projects. They devised their own ways of selecting beneficiaries, virtually losing sight of the poverty targeting requirement in the process. They favoured a random selection procedure whereby names were pulled out of a hat until the employment quota was satisfied. The communities preferred a random procedure in which every member of the community had an equal chance of being hired to a purposive targeted selection.

11.6.3 Lack of poverty definition

The targeting process for the TIP also ran into difficulties because of the lack of a clear definition of poverty (Chinsinga *et al.*, 2001; Lawson *et al.*, 2001). The criteria provided in the guidelines for VTFs (see Section 11.3) related to vulnerability. The TIP was moderately successful in targeting the vulnerable. But as vulnerability is not directly related to poverty, the focus on vulnerability in the targeting criteria increased the probability that the intervention would fail to target the poor.

The confusion between poverty and vulnerability also raises a critical question about the appropriate unit of targeting. Should households or

---

Box 11.2. The spirit of egalitarianism at Chikungulu village, Mwanza district.

The people of Chikungulu reported that the 2001/02 TIP experience was better than that of the preceding year. The main reason for the improvement was that the authorities could not agree on who would receive TIP and eventually decided that the packs should be shared equally among the clans in the village. Even at this level, it was extremely difficult to decide on beneficiaries, so they ended up dividing the contents of the packs equally between households. Participants felt that this produced less conflict than in 2000/01 because 'we have all lost'. They argued that every member of the village was poor and equally deserving of external assistance.

Source: Adapted from Chinsinga *et al.* (2002).

individuals be targeted? If vulnerable individuals are successfully targeted, the likelihood of inclusion errors increases because there may be a disproportionate share of vulnerable individuals such as orphans and elderly people in households that are generally wealthy (Lawson *et al.*, 2001).

While the emphasis of the targeting guidelines on criteria associated with vulnerable individuals helps to explain the failure of community poverty targeting under the TIP, it also suggests that the approach will have a greater chance of success in social protection projects and others which aim to target the vulnerable. Vulnerable individuals are easier for communities to identify (and therefore to select as beneficiaries) than poor households.

11.6.4 Lack of poverty data

The difficulty of targeting poor beneficiaries without reliable poverty data is also widely recognized (Johnson and Start, 2001). A fairly accurate identification of deserving beneficiaries requires an elaborate documentation of people's endowments. This is beyond the capabilities of many developing countries and is particularly difficult for a country like Malawi, where socio-economic differences between poor, rural households are insignificant (Mann, 1998).

11.6.5 Prioritizing the powerful

In the Chinsinga *et al.* (2002) study, focus group participants were asked to use the household cards to identify which households deserved to receive TIP, because they met qualifying criteria, and which did not. Three separate groups of stakeholders – members of VTFs, beneficiaries and non-beneficiaries – agreed that between 14% and 18% of households that had received TIP in 2001/02 did not satisfy any criteria for eligibility.[5] These households should not have been targeted, and represented an inclusion error.

This inclusion error was due mainly to a tendency by VTF members to give priority to themselves (self-selection) or their relatives, friends and others close to the village power structures (favouritism). Beneficiary registration was often done in secret or altered after open registration exercises that participants characterized as mere window dressing (Chinsinga *et al.*, 2002, 2004).

11.6.6 Low quotas

The study found that over half of those who did not receive a pack in 2001/02 met the communities' criteria for receiving, but were excluded. Such exclusion errors occurred mainly because in 2001/02 the beneficiary quotas were too small: the number of packs distributed – enough for roughly one-third of households – was well below the number of deserving beneficiaries, according to the communities entrusted with the targeting.

11.7 Is Targeting Possible for Free Inputs Programmes?

11.7.1 External facilitation

A pertinent question is whether community targeting is still a feasible policy option for free inputs programmes, given the problems encountered by the TIP. Some scholars and practitioners are sceptical, arguing that the informational advantage of community targeting 'may well be outweighed by an accountability disadvantage' (Ravallion, 2000). Community targeting mechanisms should be transparent, accountable and open to participation from outsiders if they are to register any measure of success.

The TIP evaluation studies found that, in Malawi, community targeting becomes more feasible if local communities are sensitized about its rationale and benefits. But both sensitization programmes and beneficiary selection need to be mediated by external facilitators. Otherwise village power structures tend to manipulate them for their own benefit. This suggests that there would be a need to involve NGOs and community-based organizations, which would imply a substantial amount of additional resources (human and financial). Targeting would become an expensive option. With reference to community-based targeting experiments in Latin America, it is argued that where targeting has been tried with minimal administrative effort and without additional resources to overcome additional costs 'it appears to have failed miserably' (Gwatkin, 2000). However, while external facilitation may be worthwhile for localized projects involving community empowerment, it seems likely to be too expensive for programmes that distribute tiny packs of agricultural inputs to large numbers of beneficiaries in thousands of villages.

11.7.2 Getting the targeting level right

The prospects for success in community targeting would be greater if the number of rural households that deserve to benefit from the programme could be accurately estimated, and the right number of packs provided (Chinsinga *et al.*, 2002). In this case, exclusion errors due to low quotas would be minimized and the outcome might be perceived to be fair.

In Malawi, in the absence of good poverty data, an alternative approach to estimating the proportion of 'deserving' households was used by Chinsinga *et al.* (2002). This study analysed the data on eligibility criteria from the household cards to find out what proportion of households should – in the view of the communities themselves – be included in the TIP and what proportion could be excluded. The stakeholder focus groups (VTFs, beneficiaries and non-beneficiaries) agreed that around two-thirds (64–68%) of households in rural Malawi should be targeted. However, this estimate had to be adjusted to allow for inclusion errors in the range of 4–8% arising from (inevitable) cases of self-selection and favouritism. This brought the total to around three-quarters of all households. Further analysis showed that the stakeholders disagreed on 30% of specific cases, but there was agreement about 20% of households which could be excluded because they did not meet any criteria for selection as beneficiaries. Thus, the study concluded that the targeting process would be considered fair by all stakeholders at a targeting level of 80% of households.

11.7.3 Geographical differentiation

However, further analysis revealed major regional disparities (Chinsinga *et al.*, 2002). In the northern region, 40% of households could be excluded, but the proportion fell to 13% in the central region and 11% in the southern region. These results suggest that geographical differentiation would be a good idea: quotas of beneficiaries could be varied between regions to reflect the magnitude of poverty and need on the ground. However, this would be problematic for political reasons. Geographical targeting is noted for its adverse political effects in countries that are politically fragmented like Nigeria and Uganda (Olayide, 2000; Hickey, 2003). Malawi is in a similar position: substantially different quota allocations for different regions would not be an attractive policy option because the government would risk losing political support. The fact is that in Malawi, as in many developing countries, voters' decisions remain linked to the anticipation of direct communal or personal benefits.

Also, there are variations within the regions (see Table 11.4). These would be likely to give rise to allegations of unfairness at the local level even if coverage was adjusted to take regional variations fully into account (Chinsinga *et al.*, 2002).

11.8 Conclusion

The 2000/01 and 2001/02 TIPs attempted narrow poverty targeting of free inputs using community targeting mechanisms. This was unsuccessful in the Malawi context. The process was hampered by factors such as the reluctance of communities to distinguish the poor; the spirit of egalitarianism in rural areas; lack of clear definitions of poverty and reliable poverty data; self-selection or favouritism by village-level authorities; and low beneficiary quotas.

The findings of the evaluation suggest that community targeting would be feasible if substantial resources were allocated to sensitization and external facilitation of community targeting processes. A successful outcome would be helped by good estimates of appropriate targeting levels, so that exclusion errors could be minimized and the outcome perceived as fair by the beneficiary communities.

However, in the case of Malawi, narrow poverty targeting does not appear to be the best option for large-scale free inputs programmes. If the right narrow targeting level is 80%, the cost of nationwide external facilitation to achieve a fair outcome would undoubtedly be greater than the cost of providing inputs for the remaining 20% of households, implying a return to

Practical and Policy Dilemmas 151

Table 11.4. Variability in the proportion of deserving beneficiaries across study sites.

Village	Region	% should receive according to VTF	% should receive according to beneficiaries	% should receive according to non-beneficiaries
Chilarika II	North	32.4	36.8	33.8
Kafwala	North	51.2	78.0	82.9
M.Chivunga	North	63.3	45.6	55.0
Mdambazuka	North	43.2	40.5	40.5
S. Chipeta	North	51.4	59.5	51.4
Chatowa	Centre	100.0	90.0	85.0
Daudi	Centre	100.0	100.0	93.3
Matapa	Centre	53.5	55.6	69.7
Mdala	Centre	81.0	77.6	67.2
Mkanile	Centre	77.9	83.8	82.4
Mkhomo	Centre	96.0	100.0	–
Nkhafi	Centre	87.1	93.5	95.2
Chimwaza	South	–	51.8	46.4
Chintengo	South	84.5	50.0	43.1
Chisalanga	South	–	89.8	95.9
Makuta	South	78.4	40.5	100.0
Mbepula	South	78.8	88.2	82.4
Mwambeni	South	74.5	60.1	57.5
Njuzi	South	62.7	41.8	83.6
Sitima	South	54.5	72.7	–

Source: Adapted from Chinsinga *et al.* (2002).

the broad poverty targeting approach. In addition, there would be the thorny political issue of variations in targeting levels between regions.

While this chapter concludes that community targeting is not appropriate for Starter Pack, the experience of the TIP evaluation indicates that it may be useful for other types of intervention. With external facilitation, the approach should be suitable for projects involving an element of community empowerment and operating in small geographical areas. Other important lessons are that there will be more chance of success if the objective of the project is to target vulnerable individuals, who are easier to identify than poor households, and if the project begins by estimating the appropriate targeting level, in order to reduce the likelihood of targeting errors.

Notes

[1] It was also specified that households should be excluded if they were benefiting from another agricultural productivity programme, had assets that could be easily converted into cash, had a regular income, or were in flood-prone areas (Chinsinga *et al.*, 2002).
[2] These are often referred to as 'qualitative studies', but strictly speaking the studies discussed here are studies using participatory methods which generated both qualitative and numerical information (see Chapter 6).

[3] Barahona and Levy (2002) set out the specific circumstances in which participatory research studies can generate statistics that can be generalized to produce valid and reliable policy results.

[4] The simulations were carried out using the household cards prepared during the 'community mapping with cards' exercise. In the Chinsinga et al. (2001) study, participants were told that there would be enough TIP packs for X number of households (one-third of the households in the village) and were asked to select this number of household cards by deciding which households most deserved to receive TIP. In Chinsinga et al. (2002), three focus groups (made up of VTF members, TIP beneficiaries and non-beneficiaries respectively) were asked to classify each household card according to whether that household deserved to receive TIP or not, and – if the household qualified – to specify the criteria by which it qualified. Codes were used to record the selection criteria on the cards; Code 10 was used for households which did not deserve TIP. It was then possible to work out what proportion of households were seen as deserving by each focus group, and for what reasons; this information could be correlated with information already on the cards about food security status and receipt of TIP (see Section 11.4).

[5] This result comes from analysing the relationship between receipt of TIP and Code 10 (see Footnote 4). Analysis of the relationship between non-receipt of TIP and the other codes allowed the researchers to establish what proportion of non-beneficiaries deserved to receive TIP.

References

Adato, M. and Haddad, L. (2002) Targeting poverty through community based works programmes: experience from South Africa. *Journal of Development Studies* 38 (3), 1–36.

Barahona, C. and Levy, S. (2002) How to generate statistics and influence policy using participatory methods in research. *Statistical Services Centre Working Paper*. University of Reading, UK. (Available on CD in Back Cover Insert.)

Barrett, C. and Clay, C. (2003) How accurate is food for work self-targeting in the presence of imperfect factor markets? Evidence from Ethiopia. *Journal of Development Studies* 39(5), 152–180.

Besley, T. and Kanbur, R. (1993) The principles of targeting. In: Lipton, M. and van der Gaag, J. (eds) *Including the Poor – Proceedings of a Symposium Organized by the World Bank and the International Food Policy Research Institute.* The World Bank, Washington, DC.

Chinsinga, B., Dzimadzi, C., Chaweza, R., Kambewa, P., Kapondamgaga, P. and Mgemezulu, O. (2001) 2000/01 TIP module 4: consultations with the poor on safety nets. (Unpublished, available on CD in Back Cover Insert.)

Chinsinga, B., Dzimadzi, C., Magalasi, M. and Mpekansambo, L. (2002) 2001/02 TIP evaluation module 2: TIP messages: beneficiary selection and community targeting, agricultural extension and health (TB and HIV/AIDS). (Unpublished, available on CD in Back Cover Insert.)

Chinsinga, B., Dulani, B. and Kayuni, H. (2004) 2003 winter TIP evaluation qualitative study. (Unpublished, available on CD in Back Cover Insert.)

Gwatkin, D. (2000) The current state of knowledge about targeting health programmes to reach the poor. (Unpublished.)

Harrigan, J. (2001) *From Dictatorship to Democracy: Economic Policy in Malawi 1964–2000.* Ashgate Publishing, London.

Hickey, S. (2003) The politics of staying poor in Uganda. *Chronic Poverty Reduction Centre Working Paper* No. 37. Institute for Development Policy and Management, University of Manchester, UK.

Hoddinott, J. (1999) *Targeting: Principles and Practice.* International Food Policy Research Institute, Washington, DC.

Johnson, C. and Start, D. (2001) Rights, claims and capture: understanding the politics of pro-poor policy. *ODI Working Paper* No.

145, Overseas Development Institute, London.

Lawson, M., Cullen, A., Sibale, B., Ligomeka, S. and Lwanda, F. (2001) 2000/01 TIP: findings of the monitoring component. (Unpublished, available on CD in Back Cover Insert.)

Levy, S. (2003) Are we targeting the poor? Lessons from Malawi. *PLA notes* No. 47. International Institute for Environment and Development, London.

Mann, C.K. (1998) Higher yields for all smallholders through 'Best Bet' technology: the surest way to restart economic growth in Malawi. *CIMMYT Network Research Results Working Paper* No.3, Harare, Zimbabwe. (Available on CD in Back Cover Insert.)

Nelson, N. and Wright, S. (1995) Participation and power. In: Nelson, N. and Wright, S. (eds) *Power and Participatory Development.* Intermediate Technology Publications, London.

Nyirongo, C.C., Gondwe, H.C.Y., Msiska, F.B.M., Mdyetseni, H.A.J. and Kamanga, F.M.C.E. (2002) 2001/02 TIP evaluation module 1: food production and security in rural Malawi (pre-harvest survey). (Unpublished, available on CD in Back Cover Insert.)

Olayide, M. (2000) Targeting development programmes in Africa: analysis of policy requirements. (Unpublished.)

Ravallion, M. (2000) *Targeted Transfers in Poor Countries: Revisiting the Trade-offs and Policy Options.* The World Bank, Washington DC.

Ravallion, M. and Wodon, Q. (1998) Evaluating a targeted social programme when placement is decentralised. *Policy Research Working Paper* No. 1945. The World Bank Development Research Group, Washington, DC.

12 Starter Pack and Sustainable Agriculture

CARLOS BARAHONA AND ELIZABETH CROMWELL

12.1 Introduction

This chapter presents the arguments and the evidence about the impact of Starter Pack on sustainable agriculture in Malawi. We first explore definitions of sustainable agriculture and examine how these relate to the Malawi situation. The relationship between Starter Pack and agricultural sustainability is then examined. Finally, we identify key issues for Starter Pack programmes and lessons for Malawi and other countries.

Why should we be concerned about sustainable agriculture in relation to Starter Pack? The smallholder farmers' focus is on immediate food security. However, they recognize that there is a need to minimize risk. Policy makers and donors also recognize the importance of ensuring that any intervention contributes to – rather than undermines – agricultural production and livelihoods in the medium to long term. Eventually, the hope is that, if sufficient attention is paid to increasing agricultural sustainability, this will contribute to poverty reduction, and interventions like Starter Pack should no longer be needed.

12.2 Sustainable Agriculture in Malawi

The term 'sustainable agriculture' is generally used to describe a broad set of principles for agricultural development, particularly those that reduce the negative social and economic impacts of 'modern' agriculture. It is also used to distinguish specific technological innovations that promote renewable or environmentally friendly methods and reduce risk.

The parameters of sustainable agriculture have grown from an original focus on ecological and environmental aspects to include economic, social and political dimensions:

- *Ecological aspects*. Reduce negative environmental externalities, enhance and utilize local ecosystem resources and preserve biodiversity.
- *Economic aspects*. Attempt to assign values to ecological parameters and include a longer time frame in economic analysis. Highlight subsidies that promote the depletion of resources or unfair competition with other production systems.
- *Socio-political aspects*. Concerned with the equity of technological change. At the local level, they emphasize farmer participation, group action and promotion of local institutions. They include attention to institutional and financial viability.

Balancing these various dimensions is one of the greatest challenges, particularly getting the balance right between short-term economic payoffs to farmers and longer-term impacts. Cromwell *et al.* (2000) found that Malawian smallholder farmers' key indicators of sustainability were closely related to ensuring immediate food security, with no reference being made to longer-term horizons or to wider ecosystem functions. Some were also found to be related to prevailing notions of best farming practices promoted by the extension services. Box 12.1 shows the five indicators of sustainability most highly ranked by smallholder farmers interviewed in 1999/2000.[1]

In our view, there are three key and inter-related concerns for sustainable agriculture in Malawi:

1. Lack of diversity in the farming system, which has negative consequences for pest and disease control, family food security and livelihoods, and agricultural biodiversity;
2. Intensity of cultivation per unit area, particularly maize cultivation, which has negative consequences for soil fertility management, food security and livelihoods at household level and, via linkages, within the rural economy as a whole; and
3. Long-term deterioration in soil fertility, which has negative consequences for intensity of cultivation and thus for food security and livelihoods.

These concerns are explored in the remainder of this section.

Box 12.1. Farmers' top five indicators of agricultural sustainability in Malawi.

Crop diversification: growing a range of staple crops
Seed availability: enough seed for timely planting at recommended spacing for all crops
Farmland size: enough land to feed the family
Tools and implements: owning all the necessary farm tools and implements
Mixed cropping: optimal mix of crops for in-field soil fertility management through intercropping and relay planting

Source: Cromwell *et al.* (2000).

12.2.1 Lack of diversity

Baseline crop and variety diversity in Malawi's smallholder sector over the last 30 years appears to have been relatively low. Although there do not seem to have been any specific investigations into crop and variety diversity, incidental evidence from other studies points to the predominance of maize in the farming system – over three-quarters of the nation's cropped area according to Smale (1991) – and, within this, to the large proportion of the area planted to local material (Smale, 1991) derived from a limited pool of landraces and old varieties including degraded hybrids (Cromwell and Zambezi, 1993).

Thus, Malawi's experience does not appear to fall within the commonly assumed paradigm of highly biodiverse small-farm agriculture at risk from the interventions of the formal seed sector. It is closer to the experience documented in Wood and Lenne (1993), of small farmers being short of crops and varieties and keenly seeking new sources.[2] Two factors may explain the situation in Malawi:

- The historical promotion of the hybrid maize and chemical fertilizer technology package and the lesser emphasis on the supply and distribution of seed and extension advice for other crops. For example, the National Seed Company of Malawi (NSCM) produced very little seed of non-maize crops for the smallholder sector during the 1980s.
- The lack of a widespread tradition of seed sharing in Malawi, compared to some other rural African societies. This may have been partly the result of the political system operating in the 1970s and 1980s, which discouraged independent, grass-roots level initiative and collaboration.

However, it appears that farmers are increasingly aware of the value of having a range of crops and income sources to ensure food security during the 'hungry period' (the months before the maize harvest). The 2001/02 food crisis highlighted for them the risks of over-reliance on maize. The Targeted Inputs Programme (TIP) evaluation surveys found that cassava was grown by 37% of farmers in Malawi in the 2002/03 season; the proportion of farmers growing sweet potatoes increased from 30% in 2000/01 to over 50% in 2002/03 (see Fig. 12.1); and rice, sorghum and bananas are also important crops in some parts of the country.

The Starter Pack/TIP evaluation programme found a clear relationship between increased food crop diversity and food security (see Fig. 12.2). Fieldwork by a non-governmental organization (NGO) network in 2004 confirmed that farmers understood the importance of crop diversity for family food security. Farmers consulted in 17 community group meetings nationwide as part of Malawi's food security and nutrition policy review process defined food security as: 'having high yields of different food crops to last the whole year . . . and availability of one or more livestock types and stable income sources'(CISANET, 2004).

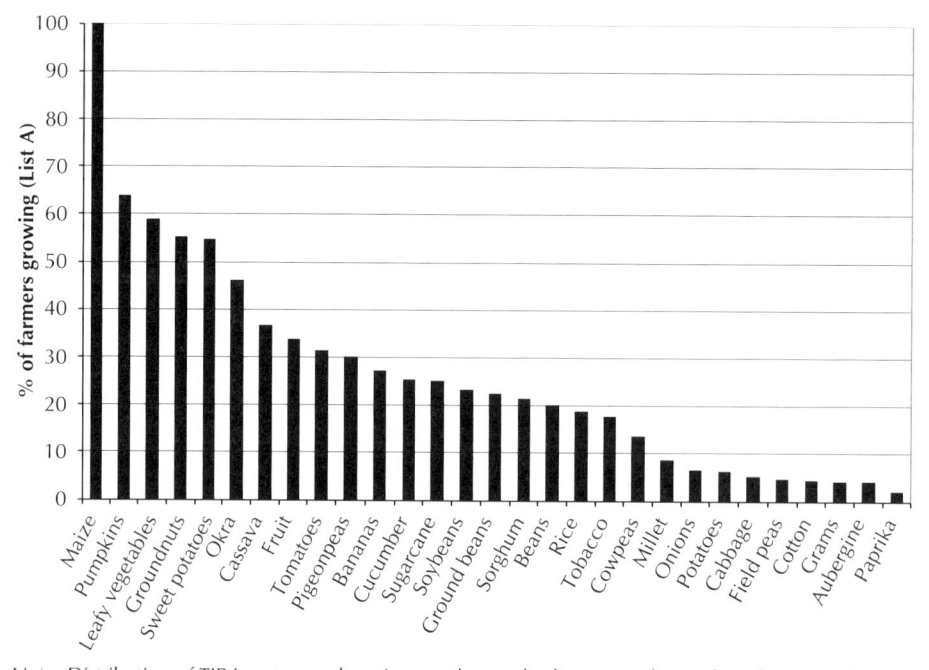

Note: Distribution of TIP inputs was done in two phases; the figures are for TIP beneficiaries who received the inputs in the first phase (List A); figures for List B beneficiaries and non-beneficiaries are similar.

Fig. 12.1. Most commonly grown crops, 2002/03. (2003 TIP evaluation survey.)

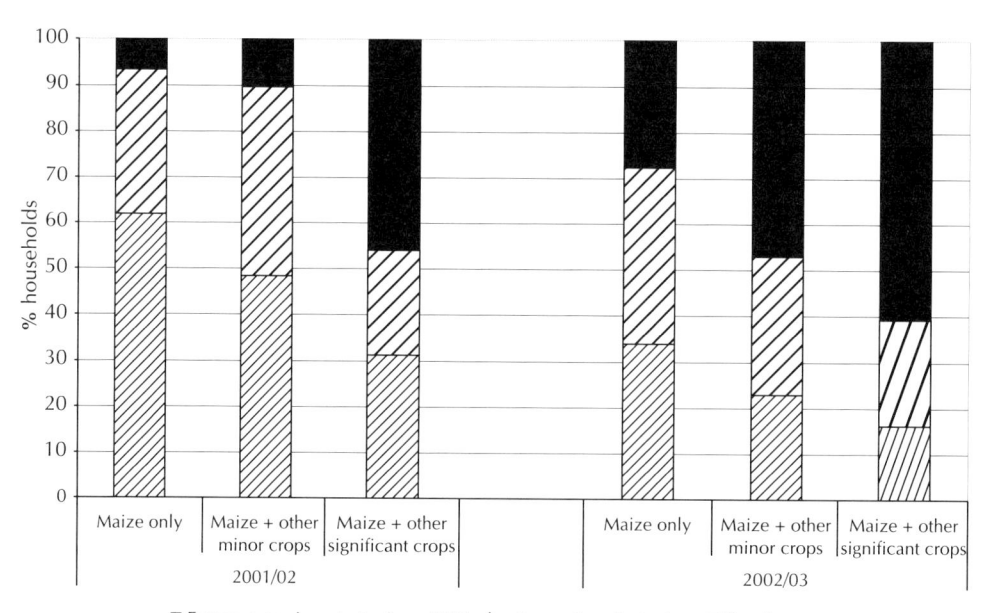

Note: To produce this graph, districts were grouped by diversity of food crops grown.

Fig. 12.2. Food security by diversity of food crops grown, 2001/02 and 2002/03. (2002 and 2003 TIP evaluation surveys.)

12.2.2 Low soil fertility

Malawi's smallholder farmers face problems of infertile soils and serious land constraints (see Chapter 9). Blackie *et al.* (1998) observed that because of increasing population density and declining land availability:

> Maize is now grown in continuous cultivation rather than as part of a fallow [rotation] which traditionally used to restore soil fertility and reduce the build up of pests and diseases. The soil resource base is now being degraded with a consequent reduction in yield.

Malawi's smallholder farmers have become highly dependent on chemical fertilizer to sustain maize yields. Sustainable agriculture technologies such as agro-forestry, integrated pest management, integrated nutrient management and conservation tillage can go some way towards reducing the need for inputs (although conservation tillage, for example, usually involves the application of herbicides). But they are demanding of both land and labour. The team that originally designed Starter Pack concluded that 'fertilizer must be an integral part of improving smallholder maize productivity in Malawi' because organic alternatives for improving soil fertility, while helpful, were insufficient in the Malawi context (see Chapter 1).

12.3 Starter Pack and Sustainable Agriculture – the Theory

Blackie *et al.* (1998) analysed how maize yields have declined in Malawi since the 1970s and attributed this change to the decline in soil fertility. The reduction in maize productivity, the increasing demand for maize from a growing population, the restrictions in land available for expanding agriculture and the liberalization of the fertilizer market were identified as key causes of the serious and chronic food insecurity at household and national levels. They proposed that the solution to the food security problem should be based on enabling smallholders to access 'Best Bet' technologies for agricultural production developed by the Maize Productivity Task Force (MPTF). According to Mann (1998) the proposed technologies were 'packages customised by region and by farmer situation. These reflect differences in soil and climate, often incorporate nitrogen fixing rotations and intercropping, and economise on costly fertilizer'.

This analysis led to the proposal of the Starter Pack programme. The proposal clearly identified the need for a long-term development effort and suggested that it should be planned for at least a 5-year period. Blackie *et al.* (1998) anticipated that as 'farmers accumulate experience with hybrid seed and fertilizer, it is reasonable to expect them to start buying small quantities on their own'.

In the longer term, the Starter Pack programme was envisaged as a means of introducing varied technologies to facilitate crop diversity. However, the programme was not conceived with the immediate objective of improving the sustainability of agriculture. The proposed 'representative pack' (see

Box 12.2) contained chemical fertilizer and hybrid maize seed. The designers of Starter Pack hoped that farmers would adopt the technology and become self-financing – but this depended on effective extension efforts and reasonable levels of input and output price stability. Both fertilizer and hybrid seeds have to be purchased each year to maintain yield, which is a risky strategy in a context of cash-poor farmers and rising input prices. A small shock can undermine purchasing capacity and/or make it unprofitable to produce maize with the recommended technology. If adoption of the technology does not happen or the inputs cannot be purchased, farmers will be unable to produce high yields.

Box 12.2. The proposed representative Starter Pack.

- 2.5 kg hybrid seed
- 10 kg 23:21:0+4S basal fertilizer
- 10 kg urea top-dressing fertilizer
- Sachet of Actellic
- Legumes (high-quality area-appropriate soybean, groundnut and pigeonpea)
- Instructional materials (leaflets)

Source: Adapted from Blackie *et al.* (1998).

However, the original proposal did include some measures which would help to promote sustainable agriculture, if implemented. It recommended crop diversification to improve food security, and crop rotations and intercropping with the appropriate legumes to improve soil fertility and nutrition (see Chapter 1). The proposal also included post-harvest technology.

As well as being a technological package of seed and fertilizer, Starter Pack was also conceived as a technology delivery system: a channel for putting the best of new agricultural technology into farmers' hands for widespread evaluation and verification, and a means of making Malawi's agricultural research and extension a two-way process. The concept and practical experience of extension is discussed in Chapter 13. However, institutional problems proved to be a major obstacle to the fulfilment of this objective, as we argue in the next section.

12.4 Starter Pack and Sustainable Agriculture – Practical Experience

12.4.1 Management and institutional issues

In practice, although Starter Pack took on board the MPTF's ideas, critical alterations were made from the first year of the programme. For instance, while the 'representative pack' proposed by the design team included 10 kg of urea for 2.5 kg of maize seed, from the outset the pack contained 5 kg of urea for 2 kg of maize seed.

Many of the changes described below were the result of decisions taken by the programme managers in response to practical problems. While it is understandable that this happened, given the programme's funding and planning constraints, it undermined the potential for contributing to agricultural sustainability and should be avoided in future Starter Pack programmes.

12.4.1.1 Simplification

Table 12.1 shows the inputs included in the packs from 1998/99 to 2003/04. The pack content varied from year to year. In 1998/99, the plan was to distribute nine different pack types, including one with cassava cuttings, which was abandoned before the distribution took place. The final distribution included five different cereal seeds and three legume seeds (Logistics Unit, 1999). In 1999/2000, the programme predominantly distributed maize seed, with a small component of rice in Karonga (Logistics Unit, 2000). From 2000/01, the programme only distributed maize as the cereal seed, because distributing alternative cereals had created logistical problems and led to complaints from beneficiaries [3] (C. Clark, Lilongwe, 2004, personal communication). The legume seed in the packs included groundnuts, soybeans, common beans and pigeonpeas. Four out of the six main-season distributions from 1998/99 to 2003/04 delivered four types of legume seed.

12.4.1.2 Uncertainty

The commitment of funds for the programme was negotiated on a year-to-year basis rather than being planned over the medium or long term (see Chapter 3). This caused uncertainty about whether the programme would go ahead each year, and prevented forward planning to obtain good quality seed of the right type. Cromwell *et al.* (2000) point out that:

> In both Starter Pack years to date, but particularly in 1999/2000, the seed included in the Pack was what was available on the regional market at the time of procurement, even though this was not the preferred or most suitable maize variety, nor the preferred legume. In fact, it was poor quality grain (not seed) in the case of a significant proportion of the legume seed supplied in 1999/2000.

Cromwell *et al.* (2000) made a series of recommendations on how to solve this problem. They argued that indicative seed requirements for Starter Pack, particularly for flint maize, should be announced to seed companies 12–15 months prior to distribution, to ensure that adequate quantities of certified seed were available for the programme. This did not happen because the programme managers were never sure of the size of the next free inputs distribution until some 3 months before the seed was required.

12.4.1.3 Inconsistency

In 1998/99 and 1999/2000, hybrid maize was the predominant cereal seed distributed, and the hybrid maize-based packs contained 10 kg of basal (23:21:0+4S) and 5 kg of urea. In 2000/01, the programme distributed open pollinated varieties (OPVs) of maize and the amount of fertilizer was reduced

Table 12.1. Inputs included in the packs.

Packs contained one cereal crop, one legume crop, basal and top dressing fertilizer. These contents could be:	1998/99	1999/2000	2000/01	2001/02	Winter 2002	2002/03*	Winter 2003	2003/04
Cereals								
Hybrid maize	●	●		●		●		●
OPV maize	●	●		●	●	●	●	●
Rice	●	●						
Millet	●	●						
Sorghum	●							
Legumes								
Groundnut	●	●	●	●		●		●
Soybeans	●		●	●		●		●
Beans	●		●	●	●	●	●	●
Pigeonpeas			●	●		●		●
Quantity of inputs provided per pack (kg)								
Cereals	2	2	2	2	2	2	2	2
Legumes	2	2	1	1	1	1	1	1
Basal fertilizer (23:21:0+4S) in pack with								
Hybrid maize or rice	10	10		5	0	5	0	5
OPV maize	5	5	5	5		5		5
Urea in pack with								
Hybrid maize or rice	5	5		5	5	5	5	5
OPV maize	5	5	5	5		5		5

Special cases: In 1998/99 if the pack contained rice the amount of seed included was 8 kg; if the pack contained millet or sorghum 0.5 kg of seed was included, but no fertilizer was provided. In 2001/02 if groundnuts were included they were unshelled and the amount was increased to 1.5 kg.

* List A distribution only.
Source: Logistics Unit, 1999, 2000, 2001, 2002 a, b, 2003 a, b, 2004.

to 5 kg of 23:21:0+4S and 5 kg of urea on the basis of technical advice from the MoA that OPVs required less fertilizer. In 2001/02, half the maize seed distributed was hybrid, and in 2002/03 almost all of the maize seed was hybrid, but the amount of fertilizer remained at the lower level of 5 kg of 23:21:0+4S and 5 kg of urea per pack, which is not the appropriate level for hybrid maize.

Extension messages were delivered through a variety of media including a leaflet in each pack with instructions about how to use the inputs. The messages were inconsistent from year to year and at some points even contradictory. Table 12.2 shows the changes in husbandry instructions from 1999/2000 to 2002/03 (the years in which the leaflets were assessed by the evaluation teams). The main reason for the changes was competition between different views within the MoA and lobbying by groups outside the Ministry (like Sasakawa Global 2000) for the promotion of technologies that they believe are appropriate for smallholder farmers. For this and other reasons (see Chapter 13), the ideal of farmer testing and adoption put forward by the designers of Starter Pack in 1998 was never achieved.

12.4.2 Content of the packs

If it were possible to solve the management and institutional problems associated with Starter Pack that gave rise to the problems outlined above, this should enhance its sustainable agriculture impact. However, there would still be a number of thorny issues relating to the type and quality of seed provided.

12.4.2.1 Type of maize seed
The debate between professionals about the type of maize seed that should be distributed in Starter Pack has been a constant feature of the programme,

Table 12.2. Husbandry instructions in the extension leaflets.

	1999/2000	2000/01	2001/02	2002/03
Distance between				
Ridges (cm)	90	75	75	Not specified
Maize seeds (cm)	25	25	25	75
Number of seeds				
per planting station	1	1	1	3
Fertilizer application				
Basal	During planting	During planting	During planting	During planting
Top dressing	3 weeks after planting	3 weeks after planting	3 weeks after planting	3 weeks after planting
Intercropping	Yes	No	No	Yes
Seed selection	No	No	No	Yes

Source: Starter Pack and TIP extension leaflets.

even before it began. Those who favour hybrids point out that the advantages of distributing hybrid seed lie in their high yield potential when coupled with fertilizer and good information. An overriding concern in Malawi is to make the most effective use of the most expensive technology component: fertilizer.

Those in favour of OPV maize seed argue that OPVs bred in Malawi, which have been released over the last decade, also have high yield potential. Elizabeth Sibale, a Malawian maize breeder who has worked in the development of both hybrid and OPV, argues that 'Although open-pollinated maize varieties cannot compete with hybrid maize in performance in high fertility environments, they do perform better than hybrids in low fertility environments' (Sibale, 2001). According to Sibale, the Malawi OPVs breeding programme aimed to develop 'a variety that farmers could afford, that was higher yielding, yet similar to local varieties in terms of storage and processing, and from which farmers could save their own next year's seed'. It is argued that Malawian OPV varieties can be recycled for 3 years without losing their productivity potential. Figure 12.3 shows the yields of hybrid and OPVs tested under farm conditions at seven sites in Malawi.

The key to resolving this debate is to understand the conditions under which smallholders grow maize in Malawi. The original designers of Starter Pack made strong assumptions about the capacity of the extension services to teach farmers the 'Best Bet' technologies (involving hybrids) and about farmers' capacity to buy inputs if they did so. If either of these assumptions fail – as was the case in Malawi[4] – then the environment remains one of low fertility and low input management systems, as indicated by Sibale (2001). Under such conditions and with promising OPVs available in Malawi, there would appear to be a case for shifting the balance towards the use of OPVs. However, more data are needed to resolve this question.

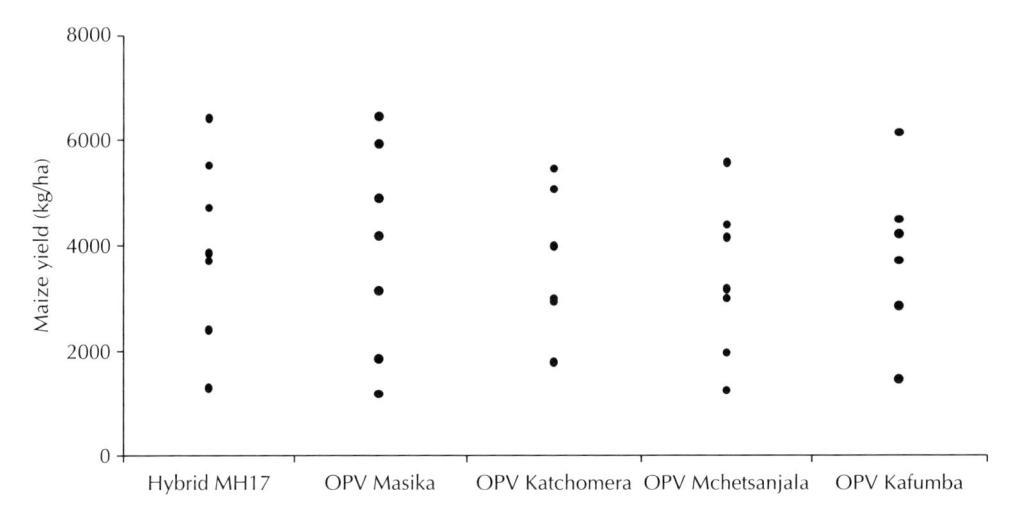

Fig. 12.3. Yield of hybrid maize and OPV under farm conditions at 7 localities throughout Malawi. (Sibale, 1995).

The Starter Pack and TIP evaluations were unable to compare the yields of hybrids and OPVs under field conditions.[5] Therefore, we cannot refute the possibility that the OPVs distributed by the programme may have been lower yielding than the hybrids (nor can we prove that this was the case). But even if OPVs produce slightly lower yields than hybrids in field conditions, they are clearly more sustainable as they reduce the small farmer's dependence on expensive inputs. If farmers learn to recycle the OPV maize seed appropriately, they will have the advantage of improved seed without having to buy it every year.

In practice, government and seed company policies significantly influence the availability of OPVs. Outright banning or official discouragement of the production and sale of OPV maize seed has affected its availability in Zambia, Zimbabwe and Malawi at one time or another. On the part of government, such actions are generally motivated by the belief that only the highest potential yield material should be made available to small farmers. Some commercial seed companies are unwilling to produce OPV seed for the same reasons and for profit considerations: the major seed purchasers in southern Africa require hybrids, and it is not profitable to produce varieties for which there is little overall demand. In this situation, the availability of OPV maize seed will depend on smallholder seed production capacity such as that of Malawi's National Smallholder Seed Producers' Association (NASSPA).

12.4.2.2 Quality of maize seed

The injection of improved seed into the diversity-poor rural smallholding was an important benefit of Starter Pack (see Table 12.3). The proportion of farmers using improved maize seed increased considerably as a consequence of the programme. There is great potential in a programme like Starter Pack to make a positive change in the quality of the seed stock among smallholder farmers. In Malawi, with an injection of around 4000 t of OPV maize seed every year reaching large numbers of small farmers, coupled with a well-run extension campaign about how to recycle the seed, there would be a strong chance of improving the maize gene pool available to farmers.

The inconsistency in the distribution of hybrid seed and OPV generated problems for the extension services. In years when hybrid seed was predominant, there was no message about seed selection; while when OPV was

Table 12.3. Use of improved maize seed.

Agricultural season	% farmers using improved maize seed	
	TIP beneficiaries	Non-beneficiaries
2000/01	87	33
2001/02	93	33
2002/03	93	26

Source: Nyirongo et al. (2001, 2002, 2003).

predominant there was a message about how to select the best seed for recycling. As most farmers cannot tell the difference between hybrids and OPVs, this probably gave the impression that the MoA was encouraging them to recycle hybrids. Most smallholders in Malawi have been doing this for many years, but the practice is not (intentionally) encouraged by the Ministry, as it leads to loss of hybrid vigour and specific vulnerabilities with certain types of hybrids.

12.4.2.3 Legume seed

The original designers of Starter Pack had clear proposals about the types of legume seed that would maximize positive impact on soil fertility as well as human nutrition. According to the 'Best Bet' research, soybeans and groundnuts were appropriate for growing in rotation, while long-duration pigeonpeas could be intercropped (see Chapter 1). However, from the outset the choice of legume was based on what seed could be procured at short notice. No attempt was made to match the legumes provided to the technology recommendations, and in the early years no extension advice was provided on use of legumes.

Some effort was made to reflect farm families' consumption preferences, but there was no effort to take into account farming practices. For instance, intercropping is widely practised in the southern region, owing to the small land areas that are cultivated, and it is unlikely that smallholders could be persuaded to change this approach. But they continued to receive legumes appropriate for growing in rotation, in terms of soil fertility benefits. Nor was there any comparison of legumes on the basis of nutritional values or potential as cash crops.

In the early stages of the evaluation programme, there was a debate between evaluation module teams about whether it was advisable to supply soybean, which has high nutritional value and good income-generating capacity but is not a preferred food in rural Malawi and requires large amounts of firewood for cooking, with implications for deforestation. The issue was never resolved, although evidence from Zimbabwe suggests that soybeans can be incorporated successfully into rural livelihoods without major changes to diet and cooking practices (M. Blackie, Bellagio, Italy, 2004, personal communication). In practice, decisions about the type of legume provided continued to be based more on logistical considerations – in particular sourcing difficulties given the short run-in time – than on any serious analysis of which types of legume should be provided.

Between 1998/99 and 2003/04, the Starter Pack programme distributed a total of 19,428 t of legumes. How important was this contribution of new seed? Nyirongo *et al.* (2002, 2003) note that many farmers grew legumes as a result of receiving the Starter Pack seed who would not otherwise have done so (see Table 12.4).

Although the pattern in Table 12.4 is consistent, the changes are modest. One possible reason is that the quantities of seed distributed for each legume varied from year to year (see Table 12.5) and this obscures the picture. Another reason is the low proportion of beneficiaries who planted the legume

Table 12.4. Growing legumes in the main season.

	2000/01		2001/02		2002/03	
	% TIP beneficiaries	% Non-beneficiaries	% TIP beneficiaries	% Non-beneficiaries	% TIP beneficiaries	% Non-beneficiaries
Groundnuts	33	38	49	39	22	13
Beans	27	24	18	12	20	15
Soybeans	20	11	10	4	23	9
Pigeonpeas	**	**	19	17	30	29
Cowpeas*	14	14	14	13	14	11

* Cowpeas were not distributed by the TIP. They are included here as an illustration of a legume crop on which the programme should have no positive effect.
** Information not recorded by the evaluation team.
Source: Adapted from Nyirongo et al. (2002, 2003).

Table 12.5. Quantity of inputs distributed by Starter Pack and the TIP.

Quantity of inputs distributed (metric t)	1998/99	1999/2000	2000/01	2001/02	winter 2002	2002/03*	winter 2003	2003/04
23:21:0+4s	27,245	29,051	7495	5002	0	9887	0	8430
Urea	14,277	14,525	7495	4995	1513	9873	1972	8407
Hybrid maize	5189	5768	0	1110	605	3408	0	0
OPV maize	485	18	2998	1017	0	464	792	3399
Rice	174	56	0	0	0	0	0	0
Sorghum	3	0	0	0	0	0	0	0
Millet	4	0	0	0	0	0	0	0
Groundnuts	3944	5765	128	487	0	886	0	1118
Soybeans	1612	0	1104	156	0	653	0	282
Beans	155	0	168	453	601	487	406	512
Pigeonpeas	0	0	53	68	0	222	0	167
Total packs distributed (millions of packs)	2.87	2.88	1.50	1.00	0.30	1.95	0.40	1.68

* List A distribution only.
Source: Logistics Unit 1999, 2000, 2001, 2002a, b, 2003 a, b, 2004.

seed provided by the programme. Often beneficiaries decided to eat the seed rather than planting it: around 30% of farmers did this each year (Nyirenda *et al.,* 2000; Nyirongo *et al.,* 2001, 2002, 2003). This reflects difficult choices between the satisfaction of immediate needs and the potential satisfaction of needs in the future. In some cases, it also reflects a rejection of the type of seed provided or late delivery of the packs.

A third factor may be the quality of the seed provided in the packs. Legume seed was reported to have had germination problems in more than one evaluation report (see, for example, Cromwell *et al.,* 2000; Nyirongo *et al.,* 2002). The Logistics Unit responded by requiring germination tests on the batches of legume seed purchased, but this did not entirely overcome quality concerns.

12.5 Other Key Issues for Starter Pack and Sustainable Agriculture

12.5.1 Promotion of organic fertilizer

Is it right for Starter Pack to promote the use of chemical fertilizer? Blackie *et al.* (1998) make a powerful case for the distribution of inorganic fertilizer, arguing that 'the decline in maize yields cannot be reversed by an organic strategy alone'. Constraints to wider use of organic soil fertility maintenance were considerable in rural Malawi in the late 1990s (see Chapter 1), and there is no evidence that the situation has improved since then. In 2001, President Bakili Muluzi launched a manure-making campaign, and since then there have been serious efforts on the part of some District Assemblies, local MoA officials and NGOs to promote manure making among smallholder farmers. But the 2003 Winter TIP evaluation found that: 'while the extension campaigns have succeeded in promoting manure awareness, they have failed in stimulating the adoption of the manure technology on a substantial scale' (Chinsinga *et al.,*2004). This is mainly because of the labour intensity of manure making, land ownership issues and problems of manure quality.

In our view, there is a case for Starter Pack to continue to supply chemical fertilizer with maize seed while the constraints on access to organic alternatives continue. Inorganic fertilizer remains essential for ensuring sufficient maize yields and meeting food security objectives. However, manure making and other organic soil fertility maintenance initiatives should be promoted in addition to the small amounts of chemical fertilizer provided by Starter Pack, as recommended in the original Starter Pack proposal.

12.5.2 Diversification away from maize

Inclusion of maize and fertilizer in Starter Pack is justified by the need to achieve an immediate increase in the availability of the main staple food in Malawi. The Starter Pack programmes of 1998/99, 1999/2000 and the Extended Targeted Inputs Programme (ETIP) programme of 2002/03, all aiming for universal or near-universal distribution, proved highly successful in achieving this (Levy, 2003). However, in the longer term, food security and nutrition in Malawi require diversification away from the current dependence on maize.

It is unrealistic to expect that maize will stop being the main staple food in Malawi, as it is rooted in the culture and taste of Malawians. There are also good reasons for farmers to grow maize where this can be done, as it is one of the most efficient plants for converting carbon dioxide and water into carbohydrates. However, a strategy for sustainable agriculture needs to encourage farmers to grow other crops that can provide carbohydrates at times when maize is not available, and crops that can be grown intercropped or in rotation with maize. It is important to increase the diversity of crops, growing not only staple food crops but a range of crops that provide food and income and allow farmers to reduce risk. As farmers themselves recognize, for achieving food security it is particularly important to grow crops that can be harvested during the hungry period or, if harvested before, that can be kept and eaten during the hungry period.

Starter Pack could play a key role in promoting agricultural diversification and thus in:

- increasing the agro-ecological sustainability of agricultural systems;
- improving human nutrition; and
- providing cash (through the sale of non-maize crops) so that families can buy food during the hungry season.

The early attempt to promote several types of cereal crop through Starter Pack foundered (see Section 12.4.1). However, conditions have changed since 1998, particularly on the logistics front. The Logistics Unit has built a system of input delivery which is now capable of dealing with different types of legume seed, and would be capable of doing the same with cereals if a sufficient lead time were allowed (see Chapter 3). In addition, Levy and Barahona (2002) propose a system of 'unpacked packs' for roots and tubers (see Box 12.3).

Box 12.3. 'Unpacked packs' for roots and tubers.

The second and third most widely cultivated staple foods in Malawi are sweet potatoes and cassava. These present problems for inclusion in Starter Packs, as they are propagated by cuttings rather than by seeds. We believe that as part of the medium term strategy for rural areas, it would be possible to promote 'unpacked packs' for roots and tubers. These would consist of plots of land in each suitable Extension Planning Area (EPA) on which good varieties of cassava and sweet potatoes would be grown. The plots would be managed by extension workers, who – when the crops were ready – would invite farmers in the area to collect samples for tasting and cuttings to plant in their fields.

Source: Adapted from Levy and Barahona (2002).

12.5.3 Time frame

For optimal impact on sustainable agriculture and food security, Starter Pack must be seen as a medium- to long-term programme planned over 5 to 10 years, not subject to re-negotiation from year to year, as has been the case in Malawi. If this were to be achieved, then a strategy could be put in place to enhance the gene pool of maize and legume crops and help in the transition towards a more diverse smallholder agriculture. In addition, the logistical challenges and opportunities faced by the programme would become more manageable and a coherent extension and communications programme could be developed.

We have argued that Starter Pack has the potential to promote sustainable agriculture. But in order to improve its performance, pack content and programme operation need to change. Adaptation of this kind would offer the potential for an eventual 'exit' from government and donor commitments to universal distribution of packs.

If Starter Pack is to become redundant in the medium to long term, it should progress:

- To include non-maize seed and 'unpacked packs'.
- From a 'one size fits all', centralized intervention to one that is tailored towards providing farmers with a range of technology options appropriate to the characteristics of their localities and that involve local stakeholders.
- From centralized extension services to a model that gives more involvement to farmers in the acquisition of information and testing and evaluation of the technologies on offer.

It may also be worth considering community production of OPV maize seed in the medium term to address problems of shortages of improved seed at local level (see Chapter 10). This would require farmer training, decentralized seed inspection, changes to seed certification regulations, the setting up of community seed banks and decisions on dissemination.

At the same time, progress has to be made on implementing a complementary growth and poverty-reduction strategy for rural areas, based on the development of smallholder farmers' livelihoods. Levy (2003) suggests that this means 'increasing opportunities for *ganyu* and other off-farm activities, promoting cash crops that smallholders can grow without displacing food crops (e.g. by intercropping), and boosting livestock ownership'.

12.6 Lessons for Malawi and Other Countries

Can we conclude that the Starter Pack programme contributes to or damages sustainable agriculture? Either interpretation would be an over-simplification. The original concept had potential for a positive impact on sustainable agriculture practices through the legume component and the ideas about intercropping and rotation. However, its focus on hybrid maize did not contribute to promoting sustainable agriculture. The introduction of OPV maize seed enhanced the programme's potential in this respect, but the extension services have so far failed to transmit the OPV maize seed selection message effectively.

If Starter Pack is to make a positive contribution to sustainable agriculture in countries like Malawi, we suggest that serious consideration needs to be given to:

- Including good quality OPV maize seed, with accompanying extension campaigns on seed selection and storage.
- Diversified pack contents.
- A carefully designed legume component.
- An increased and efficient knowledge/extension component with consistent messages on crop production, soil improvement (including organic fertilizer) and conservation.
- Implementing fertilizer recommendations that make optimum use of this expensive resource.
- A medium-term planning horizon, to allow for procurement and delivery of good quality inputs.

Would it be a good idea – from the sustainable agriculture point of view – for other countries to replicate the Starter Pack model? Here, it is important to examine the basics of the Starter Pack programme in Malawi, which are not about providing a simple 'recipe' of maize seed and fertilizer. Rather, the basic concept is about identifying constraints common to large numbers of smallholders (e.g. degraded soils and low quality seed), and assessing whether a small pack of inputs can help to overcome these constraints. If this can be achieved by a well-designed programme, the model has the potential to make an important contribution to agricultural sustainability.

Three important points should be borne in mind in any such initiatives. First, the importance of geographical scale: the intervention should reach a critical mass of smallholders, however remote, if it is to have a strong impact on diversity. Second, it needs to have a medium- to long-term time frame in order to allow procurement of appropriate seed and to make a difference to farming systems. And finally, it requires strong institutional support, particularly from the extension services. If it achieves all three objectives, it should be capable of making a contribution to sustainable agriculture and thereby supporting a country's long-term food security and poverty reduction strategy.

Notes

[1] In-depth preliminary participatory field work in three sites identified 15 indicators that farmers use to assess agricultural sustainability. In the main study, pair-wise ranking of the 15 indicators was undertaken in 30 villages to identify the five most highly ranked indicators at national level. There was some variation between regions, explained by variation in regional farming systems: for example, farmers in the central and southern regions ranked farmland size highly while those in the northern region did not. Agroforestry and the practicing of fallow were ranked low across

the country; the reasons given were land shortage, lack of knowledge and poor availability of inputs.

[2] Wood and Lenne review experience in a number of different small-farm farming systems in Africa, Asia and Latin America and conclude that 'in many traditional farming systems genetic erosion of varieties is increasing at an alarming rate' and the incorporation of new varieties into agricultural systems makes major contributions to their stability and survival.

[3] Beneficiaries complained, for instance, about receiving sorghum instead of maize. However, this was probably because sorghum packs did not contain fertilizer – a valuable commodity.

[4] Extension failure has already been discussed. Smallholder farmers' capacity to buy inputs remains very weak after several years of Starter Pack and TIP (see Chapter 10).

[5] It should have been possible to make comparisons in the one year of the programme when hybrids and OPVs were distributed in similar quantities: the 2001/02 TIP. However, the Logistics Unit was unable to provide the evaluation teams with precise information about where the different types of seed were distributed, and the teams found that farmers could not distinguish between hybrid seed and OPVs, so it was not possible to check where different types were distributed by asking farmers.

References

Blackie, M., Benson, T., Conroy, A., Gilbert, R., Kanyama-Phiri, G., Kumwenda, J., Mann, C., Mughogho, S., Phiri, A. and Waddington, S. (1998) Malawi: soil fertility issues and options – a discussion paper. MPTF/Ministry of Agriculture and Irrigation, Malawi. (Unpublished, available on CD in Back Cover Insert.)

Chinsinga, B., Dulani, B. and Kayuni, H. (2004) 2003 winter TIP evaluation qualitative study. (Unpublished, available on CD in Back Cover Insert.)

CISANET (2004) The people's voice. A community consultation report on Malawi food and nutrition security policy formulation process. Civil Society Agriculture Network, Lilongwe, Malawi.

Cromwell, E. and Zambezi, B. (1993) The performance of the seed sector in Malawi: an analysis of the influence of organisational structure. *ODI Research Study*. Overseas Development Institute, London.

Cromwell, E., Kambewa, P., Mwanza, R. and Chirwa, R. with KWERA Development Centre (2000) 1999/2000 Starter Pack evaluation module 4. The impact of Starter Pack on sustainable agriculture in Malawi. (Unpublished, available on CD in Back Cover Insert.)

Levy, S. (2003) Starter Packs and hunger crises: a briefing for policy makers on food security in Malawi. (Unpublished, available on CD in Back Cover Insert.)

Levy, S. and Barahona, C. (2002) 2001/02 TIP: main report of the evaluation programme. (Unpublished, available on CD in Back Cover Insert.)

Logistics Unit (1999) Final report: implementation of Starter Pack scheme 1998. Lilongwe, Malawi. (Unpublished, available on CD in Back Cover Insert.)

Logistics Unit (2000) 1999 Starter Pack programme final report. Lilongwe, Malawi. (Unpublished, available on CD in Back Cover Insert.)

Logistics Unit (2001) Final report: implementation of Targeted Inputs Programme 2000, Lilongwe, Malawi. (Unpublished, available on CD in Back Cover Insert.)

Logistics Unit (2002a) Final report: implementation of Targeted Inputs Programme 2001, Lilongwe, Malawi. (Unpublished, available on CD in Back Cover Insert.)

Logistics Unit (2002b) Final report: implementation of winter cropping programme 2002, Lilongwe, Malawi. (Unpublished, available on CD in Back Cover Insert.)

Logistics Unit (2003a) Final report: implementation of Targeted Inputs Programme 2002, Lilongwe, Malawi. (Unpublished, available on CD in Back Cover Insert.)

Logistics Unit (2003b) Final report: implementation of winter cropping programme 2003, Lilongwe, Malawi. (Unpublished, available on CD in Back Cover Insert.)

Logistics Unit (2004) Final report: implementation of Targeted Inputs Programme 2003, Lilongwe, Malawi. (Unpublished, available on CD in Back Cover Insert.)

Mann, C. K. (1998) Higher yields for all smallholders through 'Best Bet' technology: the surest way to restart economic growth in Malawi. CIMMYT *Network Research Results Working Paper*, No. 3. Harare, Zimbabwe. (Available on CD in Back Cover Insert.)

Nyirenda, K.F.D., Msopole, R., Gondwe, H.C.Y. and Msowoya, M.N.S. (2000) 1999/2000 Starter Pack evaluation module 2: Micro-economic impact and willingness to pay. (Unpublished, available on CD in Back Cover Insert.)

Nyirongo, C.C., Gondwe, H.C.Y., Msiska, F.B.M., Mdyetseni, H.A.J. and Kamanga, F.M.C.E. (2001) 2000/01 TIP evaluation module 2.1. A quantitative study of markets and livelihood security in rural Malawi. (Unpublished, available on CD in Back Cover Insert.)

Nyirongo, C.C., Gondwe, H.C.Y., Msiska, F.B.M., Mdyetseni, H.A.J. and Kamanga, F.M.C.E. (2002) 2001/02 TIP evaluation module 1: food production and security in rural Malawi (pre-harvest survey). (Unpublished, available on CD in Back Cover Insert.)

Nyirongo, C.C., Msika, F.B.M., Mdyetseni, H.A.J. and Kamanga, F.M.C.E. with Levy, S. (2003) Food production and security in rural Malawi (pre-harvest survey). Final report on evaluation module 1 of the 2002/03 Extended Targeted Inputs Programme (ETIP). (Unpublished, available on CD in Back Cover Insert.)

Sibale, E. (1995) Proposal to release some open pollinated maize varieties. Maize research programme, maize breeding, population improvement programme. Chitedze Research Station, Malawi.

Sibale, E. (2001) Seed security for smallholders. *Spore, information for agricultural development in ACP countries* No. 95. Available at: http://spore.cta.int/spore95/spore.pdf

Smale, M. (1991) Chimanga cha Makolo, hybrids, and composites: an analysis of farmer adoption of maize technology in Malawi, 1989-91. CIMMYT *Economics Working Paper* 91/04.

Wood, D. and Lenne, J. (1993) Dynamic management of domesticated biodiversity by farming communities. Paper presented at the United Nations Environment Programme Expert Conference on Biodiversity, 24–28 May 1993, Trondheim, Norway.

13 The Challenges of Agricultural Extension

CHRIS GARFORTH

13.1 Introduction

From the beginning of Starter Pack, it was recognized that the programme would only be successful if farmers were willing and able to use the free inputs in ways which led to an increase in their level of production. Even if everything worked efficiently to deliver the required amount of inputs at the right time to the intended recipients, farmers would have to decide if and how to use them. The composition of the packs was designed on the basis of scientific knowledge of maize and legume production in Malawi. The use of the packs on the farm would depend on farmers' knowledge and skills.

This chapter looks at the ways in which this challenge was addressed and the lessons we can draw from the experience. It begins by reviewing how ideas about extension have evolved in recent years and by describing the organization of extension in Malawi. It then describes what was done by Starter Pack to try to ensure that farmers would use the inputs effectively. We go on to review the evaluation findings with respect to extension and how they were used to reflect on and modify the extension approach and tools. The final section suggests lessons for future free inputs programmes.

13.2 Agricultural Extension

13.2.1 Changing views on extension

In Malawi, as in most countries in Africa, extension has for many years been seen as a public-sector function, of which the main objective is to encourage farmers to take up more productive and sustainable technologies. This view was premised on the belief that low productivity in agriculture is due largely to a lack of access to such technologies (such as new varieties and improved land management practices) and of the knowledge and skills needed to put them into practice, and that it was the role of government to make them available.

As understanding of and respect for farmers' knowledge and their skill in managing limited resources in situations of great risk and uncertainty has grown, the limitations of this 'transfer of technology' concept of extension have been recognized (Leeuwis, 2004). The task of extension is now more often spoken of as facilitating innovation and encouraging adaptation and experimentation than as adoption of a precise set of recommendations. Recognition of diversity has also shown the importance of farmers' learning of principles rather than simple recipes for action – and of making available options rather than a 'one-size-fits-all' technology. These changes have led to the proliferation of participatory methods of extension, in which farmers and extension staff engage in co-learning about what works under local conditions.

The assumption that extension should be a public service has also been called into question. Many private sector organizations are engaged in providing advice and information to farmers and working with them to develop solutions to local agricultural problems. Examples in Malawi include Seed Co and the National Smallholder Farmers Association of Malawi (NASFAM) (Walton, 2004). While outright privatization of government extension services is not a realistic option in most African settings, several countries are looking for ways of recouping some of the cost of these services from farmers. This is not just an economy measure: it is an important step towards greater accountability to farmers and responsiveness to what they need and want from extension.

13.2.2 Information, knowledge and new practices

The 'transfer of technology' approach to agricultural extension was based on a simplistic idea of how farmers learn new ideas. Farmers' knowledge is an important part of their human capital – one of the key assets which they use to secure their livelihoods in risky and uncertain environments. Extension advice and information interact with, and are evaluated against, what farmers already know. They may dismiss the new information, or interpret it differently from the way intended by those providing it. In the end, it is the farmer who integrates new information with his or her existing knowledge and decides whether to make any changes in farming practice.

This is why positive responses to extension advice usually take the form of adaptation rather than simple adoption of recommended practices. Farmers take from the advice what they think will work on their farms, within their constraints and their current way of doing things. Advice which suggests major changes in farming systems or practices is unlikely to be accepted from a single encounter with an extension officer or a leaflet. Similarly, it is difficult to learn new skills just from hearing or reading about them: facilitated practice is an essential ingredient in the development of all but the simplest of skills.

13.2.3 Extension reforms in Africa

From the late 1970s to the mid 1990s, public sector extension in many African countries was dominated by the Training and Visit (T&V) system promoted by the World Bank. T&V was based on the deceptively simple idea that farm productivity could be increased by the regular delivery of uniform 'messages' to large numbers of farmers. Field-level extension staff were trained each fortnight or month on the message to impart, and spent the rest of their time meeting individual contact farmers or groups to deliver the message. The system had some merit in improving discipline and planning in the management of extension services and in building stronger links between extension and research. It had shown some success in uniform environments under irrigation in Asia. However, in diverse and risky rain-fed production systems in Africa, it failed in its overall goal (Purcell and Anderson, 1998) and is now widely accepted as conceptually flawed.

Current reforms in public sector extension in Africa cover a wide range of initiatives including encouraging pluralism in supply of services, decentralizing the management and delivery of extension services, devolving responsibility for extension to local governments within a broader programme of decentralization and local government reform, cost-recovery from farmers and (as in Uganda) the contracting of private sector organizations and individuals to deliver advisory services. Central to most reforms is the objective of encouraging services to become more responsive to the needs and demands of farmers.[1]

13.2.4 Extension infrastructure in Malawi

The Starter Pack programme began in 1998 at a time of much reflection about the future of agricultural extension in Malawi, stimulated by a widespread perception that it was failing in its task of promoting the adoption of productivity-enhancing technologies (Government of Malawi, 1999, 2000).

Since 1981, the Department of Agricultural Extension and Training (later to become the Department of Agricultural Extension Services) of the Ministry of Agriculture (MoA) had been using a modified version of T&V called 'Block Extension'. Extension workers, known as Field Assistants (FAs), were allocated to an area (block) and visited groups of farmers on a regular basis to advise and train them in crop technologies and practices selected by the Ministry. During the 1990s, government progressively reduced its funding of the extension service. Staff levels fell, and by the end of the decade the

Department had experienced an 'erosion in technical expertise that, together with the financial situation, [made] the public service largely ineffective and unsustainable' (Government of Malawi, 2000). Staff morale had also fallen due to job insecurity, low remuneration and lack of resources to do the job. Any account of the extension aspects of Starter Pack must be read with this context in mind.

Malawi has come to the point of radical reform of extension later than many other countries in sub-Saharan Africa. However, the government now has a vision of a future extension system in Malawi which is pluralistic, accessible to all who want services, driven by farmer demand rather than the supply of technologies from research and managed within decentralized democratic structures (Government of Malawi, 2000).

13.3 Extension and Communication in Starter Pack

13.3.1 The message and its delivery

Technical advice on how to use the inputs provided by Starter Pack was delivered in three main ways: through a leaflet included in each pack, the FAs and radio. The leaflets were designed by the Agricultural Communication Branch (ACB) of the MoA. FAs were expected to set up on-farm demonstrations (OFDs) to show farmers how to use the inputs. Radio played the dual role of alerting farmers to the programme and reinforcing the technical advice, particularly in the later years.

In the early years of Starter Pack, the extension message was based on a project which pre-dated the programme and which identified 'Best Bet' technologies based on hybrid maize seed and fertilizer, with legumes planted separately in rotation (see Chapter 1). The technical recommendations for planting the maize and applying the fertilizer drew on the Sasakawa Global 2000 approach. The early leaflets specified that the maize was to be planted on its own, with 75 cm spacing between ridges, 25 cm spacing between planting stations and one seed per planting station (see Fig. 13.1). The technical goal was to achieve a plant density higher than farmers normally achieve, and to support this plant population through a targeted use of fertilizer at the time of planting and as a top dressing when the plants were well established. No advice was given on the use of the legume seeds until 2002/03.

As lessons were learnt from the evaluation exercises (see Section 13.3.3), a debate on the appropriate extension message took place. From 2002/03, farmers in some areas received a leaflet recommending the traditional spacing for (75 cm between planting stations and three seeds per planting station), and were encouraged to plant the legumes between maize planting stations. From winter 2003, when open pollinated variety (OPV) maize seed was widely distributed (see Chapter 12), advice was also given on how to select the best maize seed at harvest time for planting the next season.

PHUKUSI LA CHIMANGA CHA KOMPOZAITI

Unduna wa Malimidwe ndi Ulimi Wothilira

MALANGIZO AKABZALIDWE KA MBEWUYI

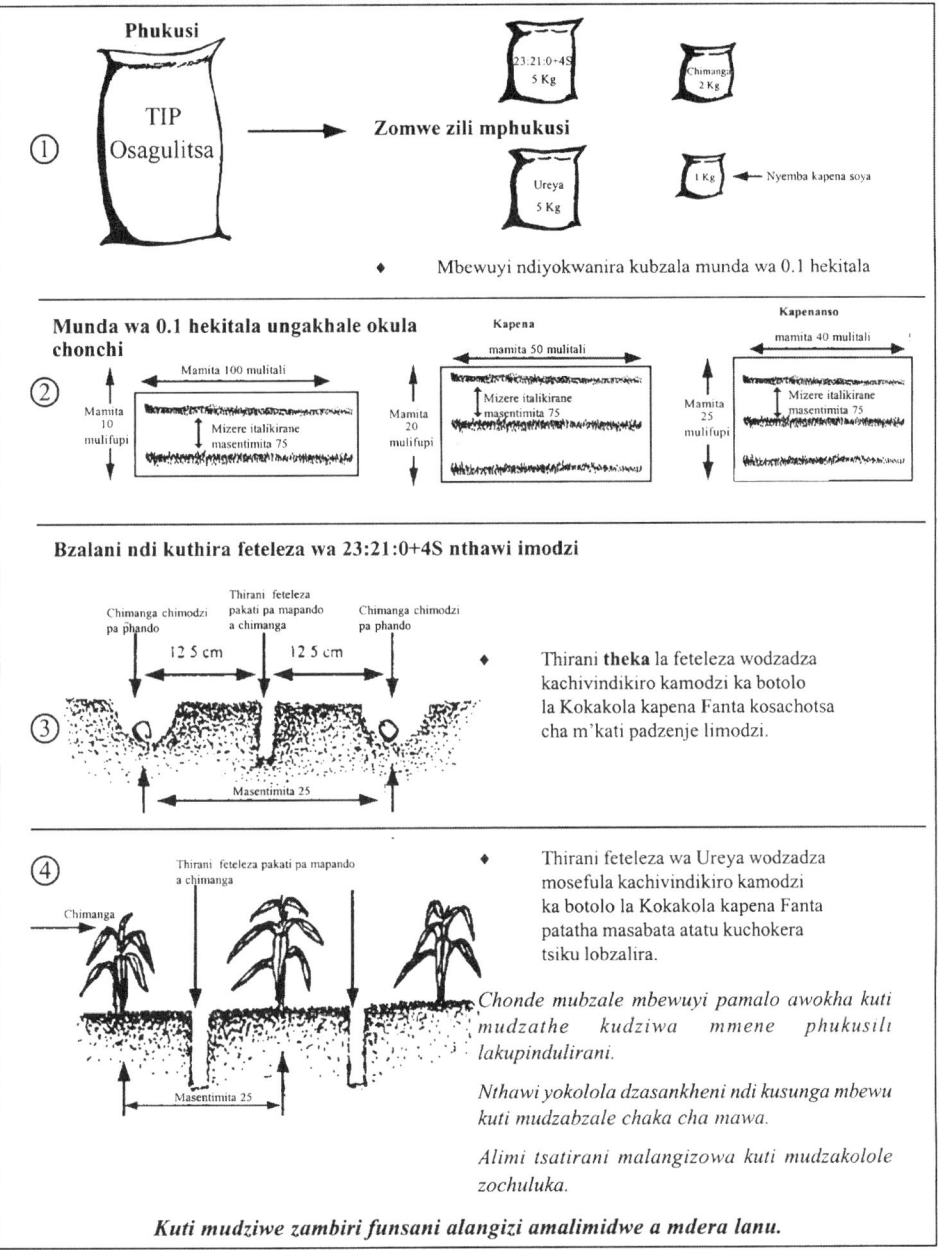

① Phukusi TIP Osagulitsa → Zomwe zili mphukusi
- 23:21:0+4S 5 Kg
- Chimanga 2 Kg
- Ureya 5 Kg
- 1 Kg ← Nyemba kapena soya

◆ Mbewuyi ndiyokwanira kubzala munda wa 0.1 hekitala

② **Munda wa 0.1 hekitala ungakhale okula chonchi**

Mamita 100 mulitali — Mamita 10 mulifupi — Mizere italikirane masentimita 75

Kapena: mamita 50 mulitali — Mamita 20 mulifupi — Mizere italikirane masentimita 75

Kapenanso: mamita 40 mulitali — Mamita 25 mulifupi — Mizere italikirane masentimita 75

Bzalani ndi kuthira feteleza wa 23:21:0+4S nthawi imodzi

③ Chimanga chimodzi pa phando — Thirani feteleza pakati pa mapando a chimanga — Chimanga chimodzi pa phando. 12.5 cm, 12.5 cm. Masentimita 25.

◆ Thirani **theka** la feteleza wodzadza kachivindikiro kamodzi ka botolo la Kokakola kapena Fanta kosachotsa cha m'kati padzenje limodzi.

④ Thirani feteleza pakati pa mapando a chimanga. Chimanga. Masentimita 25.

◆ Thirani feteleza wa Ureya wodzadza mosefula kachivindikiro kamodzi ka botolo la Kokakola kapena Fanta patatha masabata atatu kuchokera tsiku lobzalira.

Chonde mubzale mbewuyi pamalo awokha kuti mudzathe kudziwa mmene phukusili lakupindulirani.

Nthawi yokolola dzasankheni ndi kusunga mbewu kuti mudzabzale chaka cha mawa.

Alimi tsatirani malangizowa kuti mudzakolole zochuluka.

Kuti mudziwe zambiri funsani alangizi amalimidwe a mdera lanu.

Malawi Government — EU — DFID

Fig. 13.1. 2000/01 TIP leaflet. (ACB.)

13.3.2 The 2000/01 TIP evaluation

13.3.2.1 Design of the evaluation
The evaluation of the 2000/01 TIP included a study on agricultural communications. The purpose was to learn lessons for improving communication in future years. The study was designed to provide complementary insights from qualitative and quantitative methods. The evaluation team selected a sample of 27 villages to represent the diversity of socio-economic and agro-ecological conditions in the country. They collected information through separate focus group discussions (FGDs) with men and women in each village, participatory activities (including a poster-drawing exercise in which farmers gave their ideas on how to put across technical advice in ways that they would understand) and a questionnaire survey to provide quantitative data on farmers' access to information on agricultural extension and the leaflets in the TIP packs.

13.3.2.2 What was learnt about the leaflet
Less than one in five pack recipients had read the leaflet. Reasons were mixed: only 65% received a leaflet in readable condition (Sibale *et al.*, 2001); many could not read (Malawi has one of the lowest rural literacy levels in the region); some received the packs so late that the contents of the leaflet were irrelevant; and others simply did not recognize that the leaflet was there to give them advice on how to use the inputs. Those who did read the leaflet found it difficult to understand, while others rejected the advice because it was so different from what they knew: they chose to plant the maize and legumes and apply the fertilizer to fit in with their normal farming practice. Most farmers, for example, intercropped the maize with the legumes. The evaluation team concluded that 'the leaflets had not been effective in conveying the intended messages' (Dzimadzi *et al.*, 2001).

Lack of understanding of the leaflet, while partly due to low literacy and education levels, was compounded by the complexity of the visual language: symbols and standard units of measurement, for example, were not familiar to the intended audience. FGDs and the poster-drawing exercise generated lots of ideas about how the information in the leaflets could have been presented more effectively, including using feet, hands and paces as units of measurement (see Fig. 13.2), and showing people doing things rather than abstract pictures of seeds sitting in holes in the ground (see Fig. 13.3).

This is not surprising: ideas about audience participation in the design of communication media are well established in the development communication literature (Mody, 1991) and there is a lot of experience in African contexts of participatory workshops with scientists, extension workers and farmers designing extension media. The evaluation team stressed the importance, when designing communication media for agricultural and rural development, of working closely with farmers so that the designers can take proper account of their existing knowledge, attitudes and practices, and of the potential implications of the content of the media for the way they live and work.

Fig. 13.2. Using paces, hands and feet as units of measurement. (Farmers' drawings in Dzimadzi *et al.*, 2001.)

Fig. 13.3. People planting maize, applying fertilizer and selecting seed. (Farmers' drawings in Dzimadzi *et al.*, 2001.)

They also suggested that audio-visual media are needed to complement print material (Dzimadzi *et al.*, 2001).

The evaluation team recommended specific changes in the design, which would improve comprehension, and suggested that farmers should be alerted beforehand to the inclusion and the purpose of the leaflets. The recommended changes included using measurements based on the human body – for example an adult foot, rather than 25 cm, to indicate plant spacing; and a knee-high plant rather than 3 weeks after planting to indicate when to apply top dressing (Dzimadzi *et al.*, 2001). Some of these ideas were taken up in 2001/02 (see Fig. 13.4), but a thorough re-design of the leaflet did not happen until 2002/03.

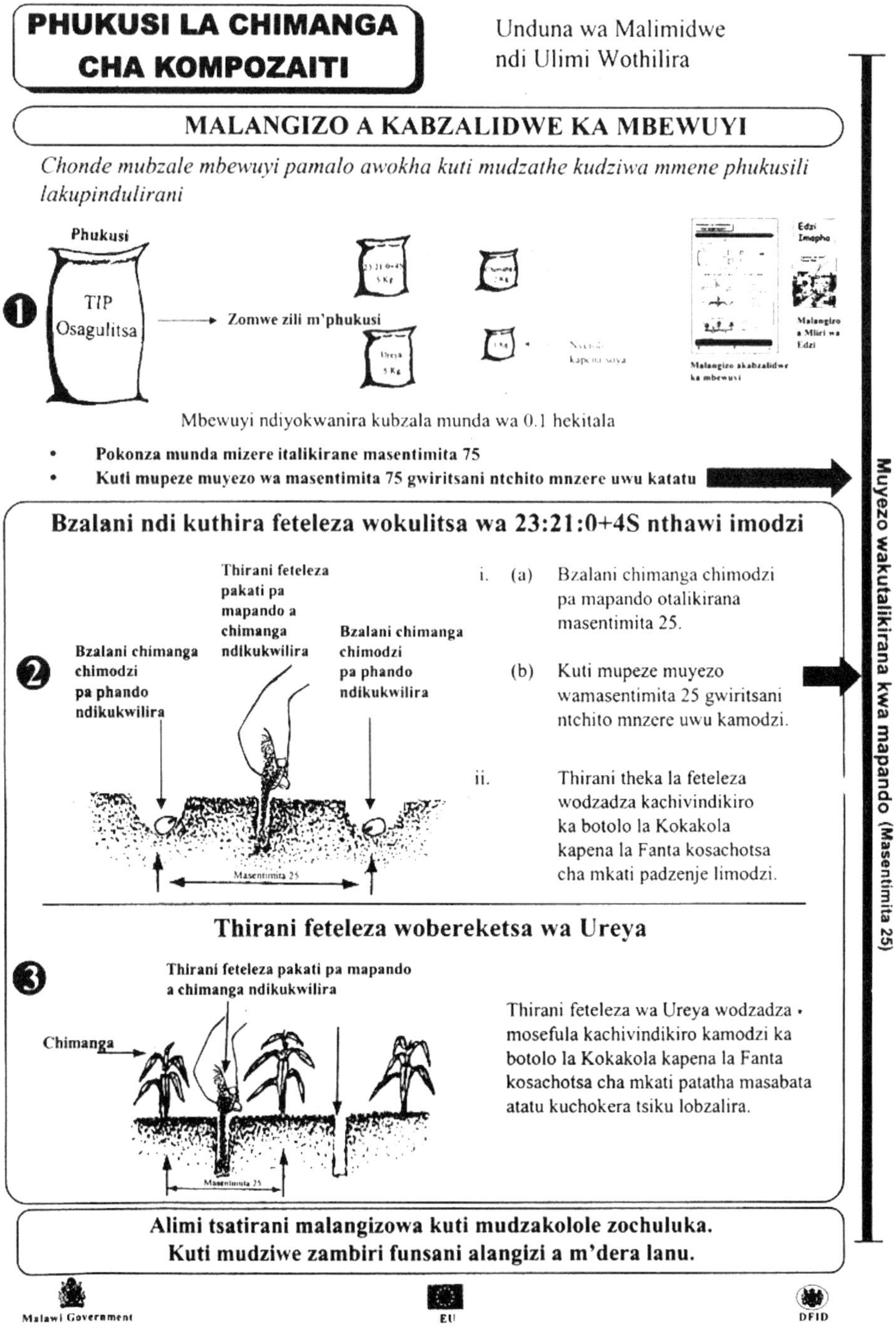

PHUKUSI LA CHIMANGA CHA KOMPOZAITI

Unduna wa Malimidwe ndi Ulimi Wothilira

MALANGIZO A KABZALIDWE KA MBEWUYI

Chonde mubzale mbewuyi pamalo awokha kuti mudzathe kudziwa mmene phukusili lakupindulirani

① Phukusi TIP Osagulitsa → Zomwe zili m'phukusi

Mbewuyi ndiyokwanira kubzala munda wa 0.1 hekitala

- Pokonza munda mizere italikirane masentimita 75
- Kuti mupeze muyezo wa masentimita 75 gwiritsani ntchito mnzere uwu katatu

Bzalani ndi kuthira feteleza wokulitsa wa 23:21:0+4S nthawi imodzi

② Bzalani chimanga chimodzi pa phando ndikukwilira. Thirani feteleza pakati pa mapando a chimanga. Bzalani chimanga chimodzi pa phando ndikukwilira.

i. (a) Bzalani chimanga chimodzi pa mapando otalikirana masentimita 25.

(b) Kuti mupeze muyezo wamasentimita 25 gwiritsani ntchito mnzere uwu kamodzi.

ii. Thirani theka la feteleza wodzadza kachivindikiro ka botolo la Kokakola kapena la Fanta kosachotsa cha mkati padzenje limodzi.

Thirani feteleza wobereketsa wa Ureya

③ Thirani feteleza pakati pa mapando a chimanga ndikukwilira

Chimanga

Thirani feteleza wa Ureya wodzadza mosefula kachivindikiro kamodzi ka botolo la Kokakola kapena la Fanta kosachotsa cha mkati patatha masabata atatu kuchokera tsiku lobzalira.

Muyezo wakutalikirana kwa mapando (Masentimita 25)

Alimi tsatirani malangizowa kuti mudzakolole zochuluka.
Kuti mudziwe zambiri funsani alangizi a m'dera lanu.

Malawi Government EU DFID

Fig. 13.4. 2001/02 TIP leaflet. (ACB.)

13.3.2.3 What was learnt about other sources of information

The Dzimadzi *et al.* (2001) study also asked farmers about their sources of agricultural information in the 2000/01 season. By far the most frequently mentioned were radio (65% of adults) and extension workers (57% of adults). More men than women said they had received information from the radio (75% and 56% respectively), while the proportions of men and women identifying extension workers as sources of information were almost identical. Scoring on the four criteria of access, relevance of content, clarity and impact during FGD discussions suggested there was not much perceived difference between radio and extension workers, and the performance of both was modest.

Farmers' access to FAs depends on whether there is one posted to the area and their commitment to their work. Many field-level posts in the MoA are unfilled, with one FA often having to serve an area that should be covered by three. Indeed, at least half of all smallholder farmers say they have never met an agricultural extension officer.[2] While acknowledging the potential value of information and advice from FAs – they felt that FAs are credible sources of relevant information because they are well trained – farmers in the Dzimadzi *et al.* (2001) study also pointed out that shortage of resources makes it difficult for them to put this advice into practice. By contrast, 'friends' were a highly accessible source of information but not always reliable because of their lack of expertise. This finding echoes those from other studies of local agricultural knowledge and information systems in Africa (e.g. Kenya: Rees *et al.*, 2000; and Eritrea: Garforth, 2001).

The results of the survey carried out by the agricultural communications study showed that 60% of men and 41% of women owned radios, but only about half of these women had the freedom to turn on the radio whenever they wanted – i.e. only one in five women had access to a radio that they could control (Dzimadzi *et al.*, 2001). Some respondents reported that the cost of batteries meant their radios were not always in working order.

Agricultural broadcasts, of which there were six each week on the national radio station in 2000/01, were among the most popular. Far fewer women than men, however, listened to these programmes. Preferred listening times are in the afternoon and evening after work in the fields has finished for the day. Understanding of agricultural information on radio varied with language, suggesting a need to have a wider range of languages on air.

Radio is a powerful tool for spreading information and awareness, particularly in Malawi, which has the highest ratio of radio sets to population in the region, while other communications infrastructure lags behind (see Table 13.1). However, in FGDs, farmers pointed out that the lack of interaction with the source of information meant there was no opportunity to clarify things they had not fully understood (Dzimadzi *et al.*, 2001). For radio to have an impact on behaviour requires that it be used in ways which stimulate discussion and interaction with peers and with other sources of information.

Table 13.1. Communication indicators for selected countries in the region.

	Adult literacy (%)	Telephones (fixed and mobile) per 1000 people	Radio sets per 1000 people	Television sets per 1000 people	Newspapers per 1000 people
Malawi	61.8	10	499	4	3
Botswana	78.9	256	150	30	27
Namibia	83.3	122	141	38	19
South Africa	86.0	364	338	152	32
Tanzania	77.1	16	406	42	4
Zambia	79.9	19	175	113	12
Zimbabwe	90.0	43	362	30	18

Source: 'ICT at a glance' tables at www.worldbank.org/data/countrydata/countrydata.html.

13.3.3 Learning from experience

13.3.3.1 The 2001/02 TIP

Based on the 2000/01 TIP evaluation, changes were made to the communication and extension component for the following year. The ACB re-designed the leaflet using some of the recommendations of the Dzimadzi et al. (2001) study, although this was not done in a participatory way. FAs were trained on the TIP recommendations and given the task of setting up 15 OFDs in each section (the area of responsibility of the FA). They were expected to help the owner of the OFD plot at critical stages (planting, fertilizer application and seed selection) and to invite neighbours to observe the recommended practices.

The 2001/02 evaluation registered some improvement. According to the evaluation survey, 83% of pack recipients received a leaflet in good condition (Nyirongo et al., 2002). However, the impact of the messages was still very low. The survey showed that fewer than one in ten farmers used the fertilizer as recommended; 57% intercropped the maize with legumes and other crops, and only a small minority followed the instructions on spacing of ridges (75 cm apart) and planting stations (25 cm apart) indicated in the leaflet. A qualitative study, which included visits to 112 TIP beneficiaries' fields (Chinsinga et al., 2002), came up with similar findings, but observed considerable regional variations, with those in the northern region (where land areas cultivated are larger and literacy rates are higher) tending to follow instructions more than in the rest of the country. Those few farmers who did follow the instructions were influenced by OFDs, or by initiatives in the area that were promoting the same technology. Farmers' reasons for not following the instructions included that the recommended approach was too labour intensive, they were not convinced it would work and they had not had any guidance from FAs. However, farmers also reported that some FAs were enthusiastic and effective, which shows what can be achieved through commitment and sensible local planning of an extension programme (see Box 13.1).

> **Box 13.1.** The Dowa success story.
>
> In 2001/02, the qualitative study team found one exceptional FA in Dowa, who, even with limited resources, managed to set up 17 OFDs. Mrs Kaonga said the key to her success was the systematic organization of her activities and building close relationships with the village heads and farmers in her section. During the TIP extension campaign, farmers turned up on time and in good numbers for Mrs Kaonga's demonstrations.
>
> Mrs Kaonga was fortunate to be located in one of Malawi's most dynamic Rural Development Projects (RDPs). In the 2003 Winter TIP evaluation, Dowa RDP was identified as a success story for its efforts to promote winter cultivation and manure making. It was also involved in crop diversification and irrigation projects. Mr Kabuluzi, the District Agricultural Development Officer (DADO), reported that the RDP had secured funding from several international donors and was also coordinating closely with NGOs.
>
> Source: Chinsinga et al. (2002, 2004).

The overall verdict of farmer participants in FGDs was that the 2001/02 leaflet was still too technical and relied too heavily on text (Chinsinga et al., 2002). FAs, on the other hand, thought the leaflet was easy to understand: in fact, some said that it gave them a clearer idea of the recommended practices than they had gained from their training.

Overall, the experience with OFDs was disappointing. The evaluation survey showed that less than 10% of TIP recipients had visited an OFD and only 18% had met an FA regularly (Nyirongo et al., 2002). Extension staff interviewed by the qualitative evaluation team claimed they had established their quota of 15 OFDs per section, but the team found demonstrations at only four out of 21 randomly selected sites. At only one site visited, was the OFD set up and used as intended. Reasons for not establishing them included:

- farmers were unwilling to volunteer their land for OFDs;
- distribution of TIP in the north was late; and
- FAs could not move around their area because they had not received their bicycle allowance.

FAs also reported some confusion over different technologies being promoted by different projects and programmes, and the fact that they feel little motivation to carry out their commitments to TIP because of the better incentives offered by NGOs to collaborate in their field activities. The evaluation team's overall conclusion was that the 2001/02 TIP extension campaign 'clearly fell short of its targets' (Chinsinga et al., 2002).

13.3.3.2 The 2002/03 TIP

This prompted the evaluation programme managers to organize a 2-day workshop on the extension aspects of TIP in July 2002. Participants reviewed the findings of the two previous seasons' evaluations and designed an improved leaflet with eye-catching cartoon-style illustrations to put across clearly articulated messages on planting maize and legumes, proper application of fertilizer, and seed selection at harvest time. The idea was to tell a story of how a farm family used the inputs and what they produced as a result. The workshop also initiated the planning of a more integrated communications campaign for 2002/03, involving the coordinated use of leaflets, radio and OFDs (Levy, 2002).

For the 2002/03 TIP distribution, packs contained a new, highly illustrated leaflet based on the one designed at the July 2002 workshop and there was more information on radio. The message also changed: programme managers had recognized that the traditional spacing for maize and the practice of intercropping legumes with maize are rational practices in many parts of the country, and this approach was included in the advice given in the leaflet. However, because of debate within the MoA, two different extension messages were distributed in 2002/03: the Chichewa version of the leaflet (for the southern and central regions) was based on the new approach, while the Tumbuka version (for the northern region) carried the same technical advice as in previous years.

The 2002/03 evaluation survey showed encouraging results on the 'reach' of the extension component. More than half of those interviewed had heard TIP messages on the radio, and the new leaflets probably had a 'moderate, positive impact' on farmers' approach to planting the maize seed. However, the extension messages had little impact on how fertilizer was applied: over 40% of beneficiaries continued to mix the basal and top dressing fertilizers and apply them together, generally quite late, while only a small minority of those who applied them separately did so at the recommended time (Nyirongo *et al.*, 2003). Farmers still reported only limited contact with FAs and OFDs.

13.3.3.3 The Winter TIP

In response to the poor food security situation, additional TIP packs were distributed for the 2002 and 2003 winter seasons (see Chapter 14). In winter 2003, the packs contained a cartoon-style leaflet, with messages expressed in dialogue between farmers and technical information shown through farmers placing fertilizer and seed (see Fig. 13.5). The leaflet also encouraged use of manure and gave advice on selection of maize seed at harvest time for planting in the next main season. Unfortunately, the evaluation team found little evidence of field extension activity to reinforce these messages (Chinsinga *et al.*, 2004).

The seed selection message was particularly important for the future of the OPV maize strategy, towards which the free inputs programme was moving, since the success of the strategy depended on farmers' ability to select good seed for recycling (see Chapter 12). Therefore, the evaluation paid particular attention to the impact of this message. However, the survey found that three-quarters of the beneficiaries of the 2003 Winter

Fig. 13.5. 2003 Winter TIP leaflet. (Montgomery Thunde (graphic artist) and ACB.)

TIP did not understand the seed selection message (Levy *et al.*, 2004). In the qualitative study, interviews were conducted with beneficiaries, who were:

> ... shown Section 8 of the leaflet, which gave instructions about seed selection and recycling, and asked what they understood from it. The most recurrent interpretation was that a family was harvesting its maize field. The family, especially the father, is very happy because they have a bumper harvest and, therefore, they are not going to worry about food for the rest of the year. The idea of seed selection and recycling was not immediately picked up by most farmers except when they were literate (Chinsinga *et al.*, 2004).

13.4 Conclusions

A lot of thought went into designing the communication aspects of Starter Pack and TIP. However, when measured in terms of the numbers of farmers using the free inputs as intended, the level of impact was well below expectations. While some improvement took place in later years, fundamental problems continued. These can be summarized under four headings: the 'message' focus; the resourcing and motivation of extension staff; lack of complementarity in communication channels; and poor implementation.

13.4.1 Extension messages

The basic extension approach was to promote a specific set of instructions and to see success in terms of the number of farmers who followed them. Particularly in the early years, this ignored the reality of farmers' existing practices based on their experience of what works in their own agro-ecological and socio-economic circumstances. Telling farmers to do things differently when they have no reason to think that their current practices are not optimal is unlikely to be effective. The fact that the instructions were not consistent from year to year can have done little to enhance their credibility among farmers.

The extension campaigns failed in their own terms by not establishing enough OFDs to provide a forum for FAs to engage farmers in a dialogue about why using the inputs in a different way from their normal practice might lead to higher production. They also failed in terms of current extension theory and practice by not encouraging adaptation or experimentation.

13.4.2 Resourcing extension staff

A recurrent problem was the low level of contact between FAs and farmers. Shortages of staff at field level were compounded by low levels of motivation and conflicting incentives and priorities. Farmers recognize the value of face-to-face interaction with trained extension staff, but a majority consistently reported that they rarely, if ever, met an extension officer. FAs, on the other hand, found that the TIP instructions differed from what they were recommending in other projects and programmes and complained of not receiving the allowances that were intended to pay for travel around their sections. This has been a common problem where governments have organized public extension systems along T&V lines with donor funding and have then been unable to resource them when donor projects have come to an end.

13.4.3 Communication channels

An important lesson from the TIP experience is that print material, however well designed and tested, will not on its own be able to convince farmers to make complex changes in their farming practice. Print material is excellent for reference, particularly for technical details which are needed infrequently and are hard to remember, such as quantities and distances. To challenge existing beliefs about the best way of using inputs, however, requires giving farmers the opportunity to question, to observe and to make their own judgements. A more coordinated and integrated approach through radio, print and FAs who are briefed to engage in dialogue and encourage local adaptation, in which each communication channel plays to its strengths, will have more of a chance.

13.4.4 Poor implementation

Key elements in the TIP and accompanying extension campaigns were not always implemented as planned. OFDs were never established in large enough numbers to play their expected role in the extension campaign. In some cases, leaflets were missing from packs or arrived in an unusable condition. There were delays in distribution, particularly in 2000/01, which gave little time for extension activities to swing into action.

On a positive note, however, there were pockets of excellent extension work, where committed staff gained the trust of farmers to establish OFDs and then used them to explain the TIP recommendations to farmers, answer their questions and generally give them the support and encouragement they needed. Where systems do not work properly, such pockets rely on the commitment and personal qualities of individual staff. The challenge for future free inputs programmes will be to ensure that effective systems are in place to encourage and enable such commitment among all extension staff.

Notes

[1] For examples, see the case studies presented at a workshop on 'New Approaches to Extension' in Washington in November 2002 (Rivera and Alex, 2004).
[2] The 2000/01 TIP evaluation survey reported 61% of survey respondents saying they 'have never met' an agricultural extension officer (Sibale et al., 2001), while the 2001/02 survey reported 57% and the 2002/03 survey reported around 50% (Nyirongo et al., 2002, 2003). This is a lower level of contact than that implied by the figures from the Dzimadzi et al., (2001) study in the previous

paragraph; but the questions asked in the two surveys were different. There is apparently a slight improvement in level of contact over the 3-year period covered by the surveys.

References

Chinsinga, B., Dzimadzi, C., Magalasi, M. and Mpekansambo, L. (2002) 2001/02 TIP evaluation module 2. TIP messages: beneficiary selection and community targeting, agricultural extension and health (TB and HIV/AIDS). (Unpublished, available on CD in Back Cover Insert.)

Chinsinga, B., Dulani, B. and Kayuni, H. (2004) 2003 winter TIP evaluation qualitative study. (Unpublished, available on CD in Back Cover Insert.)

Dzimadzi, C., Chinsinga, B., Chaweza, R. and Kambewa, P. (2001) 2000/01 TIP evaluation module 3: agricultural communications. (Unpublished, available on CD in Back Cover Insert.)

Garforth, C. (2001) Equipping the mediators: enabling extension staff in Eritrea to mediate between users and providers of agricultural information. *IAALD Quarterly Bulletin* 46 (3/4), 64–75.

Government of Malawi (1999) Review of Malawi agricultural policies and strategies. Ministry of Agriculture and Irrigation, Lilongwe, Malawi.

Government of Malawi (2000) Agricultural extension in the new millennium: towards pluralistic and demand-driven services in Malawi. Policy document. Ministry of Agriculture and Irrigation, Department of Agricultural Extension Services, Lilongwe, Malawi.

Leeuwis, C. (with contributions from A. van den Ban) (2004) *Communication for Rural Innovation: Rethinking Agricultural Extension*, 3rd edn. Blackwell Publishing, Oxford, UK.

Levy, S. (2002) Report of the Starter Pack/TIP leaflet and agricultural extension workshop, 3–4 July 2002. Report for DFID–Malawi. (Unpublished.)

Levy, S., Nyirongo, C.C., Gondwe, H.C.Y., Mdyetseni, H.A.J., Kamanga, F.M.C.E., Msopole, R. and Barahona, C. (2004) Malawi 2003 winter TIP: a quantitative evaluation report. (Unpublished, available on CD in Back Cover Insert.)

Mody, B. (1991) *Designing Messages for Development Communication: an Audience Participation-based Approach*. Sage, Newbury Park, USA.

Nyirongo, C.C., Gondwe, H.C.Y., Msika, F.B.M., Mdyetseni, H.A.J. and Kamanga, F.M.C.E. (2002) 2001/02 TIP evaluation module 1: food production and security in rural Malawi (pre-harvest survey). (Unpublished, available on CD in Back Cover Insert.)

Nyirongo, C.C., Msika, F.B.M., Mdyetseni, H.A.J. and Kamanga, F.M.C.E. with Levy, S. (2003) Food production and security in rural Malawi (pre-harvest survey). Final report on evaluation module 1 of the 2002/03 Extended Targeted Inputs Programme (ETIP). (Unpublished, available on CD in Back Cover Insert.)

Purcell, D.L. and Anderson, J.R. (1998) Agricultural extension and research: achievements and problems in national systems. World Bank Operations Evaluations Study. World Bank, Washington, DC.

Rees, D., Momanyi, M., Wekundah, J., Ndungu, F., Odondi, J., Oyure, A.O., Andima, D., Kamau, M., Ndubi, J., Musembi, F., Mwaura, L. and Joldersma, R. (2000) Agricultural knowledge and information systems in Kenya – implication for technology dissemination and development. *AgREN Network Paper* 107. Agricultural Research and Extension Network. Overseas Development Institute, London.

Rivera, W. and Alex, G. (eds) (2004) *Extension Reform for Rural Development – Case Studies of International Initiatives* (5 vols.) World Bank, Washington, DC. Available at: http://Inweb18.worldbank.org/ESSD/ardext.nsf/11ByDocName/Publications ExtensionReformfor RuralDevelopment

Sibale, P.K., Chirembo, A.M., Saka, A.R. and Lungu, V.O. (2001) 2000/01 TIP module 1: food production and security. (Unpublished, available on CD in Back Cover Insert.)

Walton, J. (2004) National Smallholder Farmers' Association of Malawi (NASFAM). In: Rivera, W. and Alex, G. (eds) *Extension Reform for Rural Development – Case Studies of International Initiatives* (5 vols.) World Bank, Washington, DC. Available at: http://Inweb18.worldbank.org/ESSD/ardext.nsf/11BDocName/PublicationsExtensionReformforRuralDevelopment

14 Why Free Inputs Failed in the Winter Season

HIESTER GONDWE

14.1 Introduction

At the peak of the 2002 hungry period in February–March, nearly 50% of smallholder farm households in the southern part of Malawi and around 60% in the central region of the country were extremely food insecure (Levy and Barahona, 2002). The situation was characterized as a hunger crisis because of the large number of deaths from a combination of disease and hunger in rural areas. The Malawi Government introduced a Winter Targeted Inputs Programme (Winter TIP) in 2002, after this crisis, in order to reduce food insecurity in the following hungry period. The programme was repeated in the winter of 2003.[1]

This chapter looks at the experience of the 2003 Winter TIP, which provided free inputs to 400,000 beneficiaries in the 'winter' (dry) season of 2003. The packs distributed in 2003 contained 5 kg of urea, 2 kg of open pollinated variety (OPV) maize seed and 1 kg of bean seed.[2] The beneficiaries were to be smallholders with access to land suitable for winter cultivation: *dambo* (wetland) or irrigated land. Following the failure to target the poor in the 2000/01 and 2001/02 main season TIPs (see Chapter 11), poverty targeting was not attempted. The packs were distributed by the end of May, in good time for winter season planting.

The policy makers' and donors' principal reason for implementing the Winter TIPs was to increase maize production in order to reduce food insecurity. They assumed that the production impact of a winter season programme would be similar to that of a main season programme. If this were the case, the 2003 Winter TIP would, with 400,000 beneficiaries, contribute between 50,000 t and 60,000 t of maize.

However, the maize production impact of the 2003 Winter TIP was disappointing. Levy *et al.* (2004) used survey data to calculate the amount of maize attributable to the programme. They observed: 'Our calculations show that – in the 'best possible' scenario – 2003 Winter TIP contributed 18,200 t of additional maize (13,000 t of grain maize and 5200 t of green maize) to national production'. In terms of household maize self-sufficiency, instead of contributing some two and a half to three 50-kg bags of maize per beneficiary household, as in the main season, the programme provided, at best, an extra half-month of maize per household.

This chapter examines why the impact of free inputs was much less in the winter season than in the main agricultural season. It shows that there is a need for careful assessment of farming conditions before a decision is taken about implementing a free inputs programme. By revealing some of the conditions in which Starter Pack is *not* effective, the 2003 Winter TIP evaluation provides pointers about which factors to check for when deciding whether a free inputs programme will be worthwhile. This is of particular importance in contexts other than Malawi, where Starter Pack has not been tested.

14.2 Winter Season Maize Farming

14.2.1 Growing maize

While almost all smallholder farm households grow maize in the main agricultural season, the evaluation survey found that under normal conditions (without free inputs), only some 70% of smallholders with *dambo* grew maize in winter 2003 (Levy *et al.*, 2004). It is therefore not surprising that one-quarter of beneficiaries of the 2003 Winter TIP decided not to plant the free maize seed in the winter season, and one-third stored the free fertilizer for later use – presumably in the 2003/04 main season. However, these decisions reduced the production impact of the Winter TIP.

Winter season farming in Malawi starts in May. When implementing the Winter TIP, the Logistics Unit worked on the assumption that winter maize is generally planted in May/June and harvested in September/October. However, the 2003 Winter TIP evaluation survey found that more than half of all farmers who planted winter maize in 2003 planted it in August/September and expected to harvest between December and February 2004. Most of those who planted late apparently did this so that part of crop growth would coincide with the summer rains.

Figure 14.1 shows two peak maize planting times: June and September. Although there is some relationship between the time of receiving packs and the time of planting maize, this may be because in some areas there was perceived to be less of a rush to get packs to farmers, given that planting was not imminent (Levy *et al.*, 2004).

Why do many farmers in Malawi not plant maize in the winter season, or plant it right at the end of the dry period? The reasons are to be found partly in the nature of *dambo* cultivation, which is discussed in the remainder of this section, and partly in the constraints of winter season farming, which are discussed in the next section.

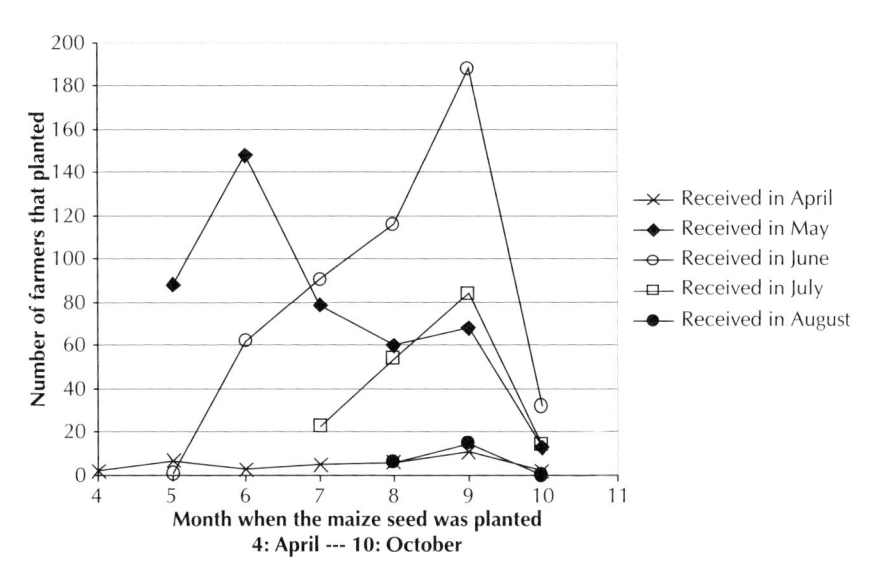

Fig. 14.1. Planting of Winter TIP maize seed. (Carlos Barahona using data from 2003 Winter TIP evaluation survey.)

14.2.2 *Dambo* cultivation

There is controversy over the definition of *dambo* and *dimba* (Chinsinga *et al.*, 2004). This chapter uses the two words interchangeably, but makes a distinction according to the type of water source available: land that is flooded (often waterlogged) in winter, and land where winter cultivation depends on residual moisture in areas bordering streams and rivers. Soils tend to be more fertile in the first category, but both categories suffer from lack of predictability. Agronomic conditions vary from year to year depending on levels of flooding (in the first category) and drying out of residual moisture (in the second category). In areas where the availability of water is problematic, winter planting, particularly of maize, is delayed until near the start of the rainy season to shorten the period of irrigation.

Differences in agronomic conditions determine the type of crop to be planted. The distribution of maize in the Winter TIP packs assumed that it does well in *dambos*. However, Chinsinga *et al.* (2004) found that low-lying *dambo* land that is subject to flooding – for instance in Karonga and Nsanje districts – is not suited for winter maize cultivation as it is often waterlogged during the winter season.

A related factor is the fertility of the soil. The evaluation studies found that uptake of Winter TIP was generally high in *dambos* that are less fertile (the second category), but low in the more fertile *dambos*, especially where alluvial sediment from flooding rivers is deposited (see Box 14.1). In such cases, the free fertilizer was kept for use in the main season. Where both *dambos* and main season gardens are considered fertile and the main season crop does not require fertilizer, the Winter TIP fertilizer was simply sold.

The Chinsinga *et al.* (2004) study concluded that:

> ... uptake of free maize seed and fertilizer for winter season cultivation are likely to be strong only where maize is traditionally grown as a *dimba* crop and the soil is not naturally fertile. Of our fourteen sites, maize was a traditional crop in ten sites, of which only seven required fertilizer to boost soil fertility. Thus, in half of the sites visited by the study, it will continue to be rational behaviour for farmers to decide against using the Winter TIP maize and fertilizer in the winter season.

14.3   The Constraints in Winter Season Farming

14.3.1   The input constraint

In the main agricultural season in Malawi, smallholder farmers' key constraints on increasing maize production are a combination of:

- Low and declining soil fertility, leading to low yields (see Chapter 1);
- The rising price of chemical fertilizer, making it unaffordable for the majority of poor, smallholder farmers (see Introduction and Chapter 10).

The combination of these factors amounts to a severe input constraint, as farmers are heavily dependent on fertilizer to address the problem of soil fertility and increase their maize yields, but they cannot afford to buy it. Other constraints such as small landholding sizes and lack of labour, while also important, are not binding in most cases. In other words, the key factor that prevents smallholders from increasing output of maize is lack of access to inputs. This is the reason why tiny packs of free inputs have a big impact.

Box 14.1. Effect of agronomic conditions on use of inputs.

The household is male-headed and has 1.5 acres of *dambo* land, located only a few metres from the house. However, not all of the *dambo* is being cultivated because part of it is waterlogged. On the part that has been cultivated, sweet potatoes have been planted.

The household has never used either manure or fertilizer on the *dimba*. This is because the household head's view is *'charo chiri makora waka'* meaning 'the soils are naturally fertile'.
Beneficiary of Winter TIP, James, Karonga.

This household, which is female-headed, has a piece of *dambo*, which is close to one acre. The household head has not cultivated the entire garden this year because part of it is submerged. She is growing sweet potatoes and rice.

She indicated she has never used fertilizer on her *dimba* garden. She does not use manure either. The soils in her area are believed to be very fertile. She pointed out: *'Ife feteleza ndi wa manyowa amtibweletsera in anthu aku Thyolo ndi Mulanje kudzera mu Ruo. Ruo akabwera sibasi talandira manyowa'* meaning 'We get our fertilizer and manure from people of Thyolo and Mulanje via Ruo River. The Ruo brings us manure.'
Non-beneficiary, Ngena, Nsanje.

Source: Chinsinga *et al.* (2004).

Although lack of fertilizer is a constraint for some smallholder farm households in the winter growing season, it is less important than in the main agricultural season. Levy *et al.* (2004) found that, in general: '*Dambo* maize cultivation takes place on better soils resulting in less need for fertilizer and producing yields per acre that are at least twice as high as in the main season.' With better soils to start with in winter, free fertilizer does not make as much of a difference to output as in the main season.[3]

14.3.2 The labour and land constraints

In winter, the main problems for smallholder farmers are shortages of labour and small land areas suitable for cultivation:

- *The labour constraint*. Chinsinga *et al.* (2004) found that: 'The most important factor limiting *dambo* use is its labour intensity. *Dambo* cultivation requires energetic people to till the soils, which usually have more clay content than upland soils. It needs determination and hard work not only during land preparation but also throughout the growing season especially where primitive methods of irrigation are involved.' Households with elderly or sick adults had particular difficulty cultivating *dambo*. Levy *et al.* (2004) found that households cultivating *dambo* in winter 2003 were larger on average than in the main season, suggesting that small households were often excluded because they did not have enough labour to engage in winter cultivation.

  Winter cultivation is also affected by the competing labour demands of other activities (farm and non-farm). Competing activities are usually those which generate cash and are considered more lucrative than *dimba* cultivation. Chinsinga *et al.* (2004) found that in Nsanje, *dimba* cultivation faces stiff competition from cotton farming and fishing. In one case in Karonga, use of the Winter TIP inputs was delayed because the household was preoccupied with cultivation of cash crops in the *dimba* to service a loan for a treadle pump.

- *The land constraint*. The 2003 Winter TIP evaluation survey established that average *dambo* landholding size per household was only one acre, of which half to three-quarters was cultivated in the 2003 winter season. This compares with an average cultivated land area per household of just over two acres in the 2002/03 main season (Nyirongo *et al.*, 2003). There are also problems of soil erosion and fragmentation of *dambo* gardens, which farmers perceive as having led to a reduction in *dambo* land areas cultivated in recent years.

These constraints contribute to reducing the maize production impact of winter free inputs programmes in Malawi. Insufficient land or labour means that the seed provided by the programme is less likely to be planted in addition to maize seed which the farmer already has; it is likely to replace seed that would have been planted anyway. Even if the farmer has access to a large area of land suitable for winter cultivation, which would allow him to expand the area of maize that he grows with the seed from a free inputs programme, he is unlikely to have enough labour to do so. In addition, the labour constraint often leads farmers to delay planting of maize until August/September, as late planting reduces the amount of labour needed for irrigation by taking advantage of the early summer rains.

14.3.3 The threat from roaming livestock

Roaming livestock is also a problem in some areas. *Dimba* cultivation, like upland cultivation, is only possible if the farmer has a reasonable guarantee that the crop will be safe. However, roaming animals threaten crops during winter season because *dimba* gardens are usually an extension of the villagers' common, which is used for grazing animals in winter. Failure to regulate the movement of animals grazing on the common may mean that winter cultivation becomes impossible (see Box 14.2).

14.4 Crop Diversity

A more basic question also needs to be asked: 'Should maize be promoted for winter cultivation?' In the case of main season cultivation in Malawi, there is little doubt about the answer. Maize is the main staple food crop; it is cultivated by almost all smallholder households; and Malawi has a serious maize deficit (see Chapter 8). However, in other contexts – even in the same country in a different season – the answer may not be as straightforward.

In addition to maize, the survey results presented by Levy *et al.* (2004) identify a number of other important *dambo* crops: leafy vegetables (grown by

---

Box 14.2. Roaming livestock.

The household, which is female-headed, has five members. The household has a *dimba* plot, which is less than a quarter of an acre. Half of this *dimba* was cultivated at the start of the 2003 winter season, where the bean seed from the Winter TIP was planted. However, as soon as the beans had germinated, livestock ate the entire crop.

While some farmers from the village erect fences around their *dimbas*, she is not in a position to do so, as this is a man's job: '*ndimasowa ondimangira phaphu, ndife mbeta*' meaning 'There is no one to [erect the fence] for me, as I am a spinster.' This experience has not only discouraged her from planting the Winter TIP maize seed, but also from engaging in winter cultivation in future: '*Kudziko kwathu kuno nkovuta ziweto. Zimenezi zinandigwetsa ulesi, basino ndasiya kulima kudimba . . .*' meaning 'We have a problem with livestock here. This has discouraged me, so much that I have given up *dambo* cultivation [in winter].'
Beneficiary of Winter TIP, Kapichila, Salima.

Source: Chinsinga *et al.* (2004).

56% of smallholders with *dambo*), pumpkins (54%), tomatoes (53%), beans (41%), sweet potatoes (35%), sugarcane (29%), tobacco (22%), European potatoes (14%), bananas (13%) and fruit (11%). Chinsinga *et al.* (2004) observed that 'in most sites there was a wider range of crops grown in the *dambo* than in upland gardens.' Since the 2002 food crisis, alternative crops to maize have become increasingly popular (see Box 14.3).

The Winter TIP raised fears that the free maize seed would displace other winter crops which diversify farmers' food and income sources. Hypothetically, this would have happened if farmers had decided to plant the maize seed provided by the programme instead of other crops that they would normally have cultivated. The Ministry of Agriculture (MoA) is committed to promoting diversification to reduce the domination of maize in the food basket and the risk of a shock to farmers' livelihoods should the maize crop fail.

The 2003 Winter TIP evaluation studies did not find evidence of displacement of other crops by maize as a result of the intervention, probably because uptake of Winter TIP maize seed occurred almost exclusively in areas where maize was already a dominant winter crop. Although this allayed fears of a negative impact on crop diversity in the case of Malawi's Winter TIP, the concern remains a valid one, which should be taken into account in the design of future interventions. If there are serious land or labour constraints, the distribution of particular types of seed may displace other crops because farmers have to choose between their traditional crops and those provided by the intervention. Care needs to be taken to ensure that free inputs programmes do not undermine crop diversity. It may even be possible to increase the non-maize component of the packs to make a positive contribution to crop diversification (see Chapter 12).

Box 14.3. Dual Role of Winter Sweet Potatoes.

G owns a *dimba* which is roughly a quarter of an acre. She has planted only sweet potatoes. She decided against maize because the *dimba* is waterlogged and she feared that maize would not do well.

She favours sweet potatoes because they can serve a dual purpose. '*Mbatata ingakupulumutse ku usiwa ndi njala*' meaning 'Sweet potatoes can protect one from both poverty and hunger.' Sweet potatoes can shield one from poverty because they can readily be sold at good prices to buy basic needs items, and they can shield one from hunger because it is now possible to make *nsima* from sweet potatoes.

G pointed out that the 2001/02 hunger crisis has taught people to be innovative. They are now able to prepare flour from sweet potatoes in the same fashion as from cassava. This has made it necessary to preserve sweet potatoes. The sweet potatoes are peeled and then sun dried and stored as '*makaka*'.
Beneficiary of Winter TIP, Ngena, Nsanje.

Source: Chinsinga *et al.* (2004).

14.5 Conclusion

In a programme designed to increase national maize production and food security, it is important to consider the nature and constraints of maize farming. The first part of this chapter discussed these issues and presented the evidence from the 2003 Winter TIP evaluation, which shows why Starter Pack is *not* effective in certain conditions.

The evidence shows that in Malawi, the maize production and food security impact of free inputs is much less in the winter season than in the main season. This is due to the nature of agronomic conditions in winter, the constraints faced by farmers and the type of crops preferred for winter season farming. Because of these factors, uptake of free fertilizer and seed is lower in winter than in the main season, and the net addition to output is less for those beneficiaries who do use the inputs. The key policy implication for Malawi is:

- Where financial resources are limited, it is best to concentrate them on a main season free inputs programme.

The second part of the chapter examined concerns about the impact of free inputs on crop diversity. In Malawi, winter season cultivation plays an important role in diversification, with relatively high proportions of households that are engaged in winter cultivation growing non-maize staple food crops and income-generating crops. The main policy implications are:

- Winter season interventions in Malawi could focus on promoting improved varieties for non-maize crops to enhance crop diversification.
- When considering whether to implement a free inputs programme in a different context, it is important to be aware that there is a risk of damage to crop diversity; programmes should always be carefully designed to minimize these risks and, if possible, to make a positive contribution to crop diversification.

Finally, a more general lesson of Malawi's Winter TIP experience is that it is important to study the specific conditions faced by beneficiary farmers carefully, to determine whether the proposed intervention is appropriate before deciding to proceed with it. While Starter Pack is very effective in some contexts – particularly where the input constraint is binding – in others it may have a low benefit to cost ratio, or might even be damaging.

Notes

[1] The 2002 Winter TIP was evaluated using a survey which raised more questions than it answered. Therefore, a full evaluation using both survey ('quantitative') and participatory ('qualitative') methods was designed for the 2003 Winter TIP. The findings presented in this chapter come from the 2003 Winter TIP evaluation reports.
[2] Basal fertilizer was not considered necessary for winter cultivation by the MoA on the grounds that soils used in winter should be rich in nutrients required for plant growth due to seasonal flood-

ing. It was also felt that farmers would use organic manure on their crops following intensive government manure utilization campaigns (Levy et al., 2004).

[3] It could be argued that if areas with less-fertile *dambo* (the second category) were targeted to receive Winter TIP, while those with fertile *dambo* (the first category) were excluded, a greater production impact might be achieved. It could also be argued that second category area farmers should receive basal as well as urea. However, such an approach implies geographical targeting, which would be difficult and politically unpopular (see Chapter 11).

References

Chinsinga, B., Dulani, B. and Kayuni, H. (2004) 2003 winter TIP evaluation qualitative study. (Unpublished, available on CD in Back Cover Insert.)

Levy, S. and Barahona, C. (2002) 2001/02 TIP: main report of the evaluation programme. (Unpublished, available on CD in Back Cover Insert.)

Levy, S., Nyirongo, C.C., Gondwe, H.C.Y., Mdyetseni, H.A.J., Kamanga, F.M.C.E., Msopole, R. and Barahona, C. (2004) Malawi 2003 winter TIP: a quantitative evaluation report. (Unpublished, available on CD in Back Cover Insert.)

Nyirongo, C.C., Msiska, F.B.M., Mdyetseni, H.A.J and Kamanga, F.M.C.E. with Levy, S. (2003) Food production and security in rural Malawi (pre-harvest survey) Final report on evaluation module 1 of the 2002/03 Extended Targeted Inputs Programme (ETIP). (Unpublished, available on CD in Back Cover Insert.)

15 Financing and Macro-economic Impact: How Does Starter Pack Compare?

SARAH LEVY

15.1 Introduction

If Malawi is to become food secure, it must close its food gap, currently estimated at 500,000 to 600,000 t of maize (see Chapter 8). Key components of a long-term strategy for rural areas should be diversification into production of non-maize food crops to complement maize production; and growth and rural development programmes to boost smallholder farmers' incomes and purchasing power. If these efforts bear fruit, the parameters of the analysis will change. However, for the next few years, non-maize crops cannot be relied upon to bridge the food gap,[1] nor do we expect a rapid reduction in income poverty. Therefore, the medium-term options are to:

- import substantial quantities of maize; or
- produce more maize.

These options are discussed in this chapter. Safety nets are also discussed because some members of Malawi's donor community have expressed the view that they constitute a food security policy option. The focus of the analysis is a comparison of the costs – direct (burden on public finances) and indirect (adverse macro-economic impact) – and benefits of each option. The analysis also takes into account risk and uncertainty. It presents evidence from the early 2000s, in particular the period of the food crisis (2001/02) and the following year (2002/03), when crisis was avoided by importing maize and organizing food aid.

We assume that none of the options is 'fiscally sustainable' in the Malawi context. The appropriate question for Starter Pack and comparable options for dealing with food insecurity is not 'Is the programme fiscally sustainable?' but 'Is it worthwhile enough to be a spending priority?' An equally important question is that of the cost of *not* implementing the intervention, which may be higher than the cost of implementing it (Levy *et al.*, 2004).

15.2 Importing Maize

Stevens *et al.* (2002) point out that underproduction of maize and large year-to-year variations in output are not new in Malawi. The country was in the same situation during the 1990s as it was in the early 2000s, but in the 1990s food shortages were overcome by importing maize. For instance, following a very poor maize harvest in 1992, imports reached 347,000 t in 1992 and 490,000 t in 1993. Again, in 1998 the country imported 325,000 t of maize following a poor harvest. Stevens *et al.* (2002) are 'puzzled' about why so few imports were ordered by the government in 2001/02, given that the dollar price of South African white maize had varied little since the early 1990s.

The answer to the puzzle of why importing maize was not an obvious option for the Malawi Government in 2001/02 is twofold, and both parts of the answer relate to the macro-economic context. The first concerns government finances and foreign reserves; the second concerns the question of consumer price subsidies. These issues are explored in the remainder of this section.

15.2.1 Paying for imports

In the early 2000s, government finances were extremely tight – the fiscal deficit reached 14% of GDP in 2000 – and the Ministry of Finance (MoF) was under intense pressure from the International Monetary Fund (IMF) to avoid additional expenditure. Moreover, foreign reserves were only US$200–300 million; with the price of South African white maize delivered to Malawi estimated at around US$245/t in 2001/02, the cost of importing enough to make an impact on food prices and food security (probably around 300,000 t) would have been some US$74 million, which would have reduced foreign reserves to dangerously low levels. Thus, the cabinet ordered only 139,000 t of South African maize at a cost of US$33 million. In 2002/03, after the seriousness of the food situation had become clear, 235,000 t of maize imports were ordered at a cost of US$75 million.

Despite efforts to disguise the impact of these maize imports on public expenditure, there were clearly fiscal implications. In Financial Year (FY) 2001/02, the US$33 million bill for imports was financed by a short-term loan which the National Food Reserve Agency (NFRA), with the MoF as guarantor, negotiated with the South African bank ABSA (Levy, 2004). The NFRA and the parastatal Agricultural Development and Marketing Corporation

(ADMARC) intended to recover some money from maize sales, but it seems that most of the funds required for the ABSA loan payments came from increases in domestic financing. One version is that the NFRA and ADMARC had a line of credit with the Reserve Bank of Malawi (RBM). Government borrowing from the RBM and from commercial banks appears to have increased rapidly from 2001, despite a sharp fall in the acknowledged fiscal deficit that year (see Table 15.1).[2]

In FY 2002/03, the commercial maize imports cost the government MK7.4 billion (4.4% of GDP).[3] There was no cost recovery that year (Levy, 2004), so the full amount had to be financed by government borrowing. In August 2002, the IMF agreed to provide a stand-by loan of SDR17.35 million (US$23 million) in emergency assistance, which was worth around MK1.8 billion when it was disbursed in September 2002. That left MK5.6 billion to be financed domestically – either by 'printing money' or by increased issues of Treasury bills.

The domestic financing mechanism appears to have combined both approaches: the government borrowed from the RBM – which partly explains the sharp increase in the monetary authorities' net credit to central government in 2002 (see Table 15.1). There was also much financing through securities, leading to an increase in Treasury bill stocks and in commercial banks' net credit to central government. This trend increased in 2003. Indeed, some of the government's debt with the RBM seems to have been securitized that year, as RBM net credit to the central government declined while commercial bank net credit and Treasury bill stocks continued to rise sharply.

The maize imports also cost Malawi dear in terms of foreign exchange reserves. These fell sharply after August 2001 (see Fig. 15.1). This was partly due to the use of foreign exchange to import maize and partly a result of the reduction in foreign aid inflows following a scandal over the government's handling of the Strategic Grain Reserve (SGR). By mid-2002, reserves had recovered somewhat – to over US $200 million; but the use of US$75 million of the RBM's foreign exchange to pay for maize imports reduced them to just

Table 15.1. Expenditure, revenue and domestic financing, 2000–2003.

	2000 (MK million)	2001 (MK million)	2002 (MK million)	2003 (MK million)
Government expenditure	32,121	31,494	45,924	60,963
Taxation and other revenue	17,771	22,853	26,515	36,356
Fiscal deficit (% of GDP)	14,350 (14%)	8640 (7%)	19,409 (13%)	24,606 (13%)
Monetary authorities' net credit to central govt	−1068	3663	9752	8155
Commercial banks' net credit to central govt	1185	2166	5960	10,449
Holdings of Treasury bills*	9817	5099	28,890	46,976

* Figures shown include interest payments.
Source: Reserve Bank of Malawi (RBM).

over US$120 million by March 2003. This led directly to exchange rate depreciation, with the Malawi Kwacha reaching MK108:US$1 at the end of August 2003, compared with MK65:US$1 2 years earlier, before the start of the food crisis.

Thus, the decision to import maize – while an understandable response – had grave macro-economic consequences. The combination of financing through domestic credit expansion and the running down of foreign reserves, leading to exchange rate depreciation, produced inflationary pressure. Substantial increases in Treasury bill stocks meant both crowding out of private sector investment (through high interest rates) and increased interest payment obligations by central government, crowding out discretionary public expenditure in future years (Whitworth, 2004).

15.2.2 Consumer price subsidies

The second part of the answer to the puzzle of why importing maize is no longer a viable strategy has to do with consumer price subsidies. When the government decided to import maize in August 2001, it calculated that the landed cost of maize would be MK14 per kg (US$220/t with an exchange rate of MK65:US$1). Thus, it fixed an ADMARC sale price of MK17 per kg for the 2001/02 season. In the event, losses occurred because the landed price rose, in both dollar and MK terms, with each shipment that arrived over the 2001/02 season.

In 2002/03, the losses rose because the scale of the commercial maize imports increased and the commitment to sell at affordable prices was main-

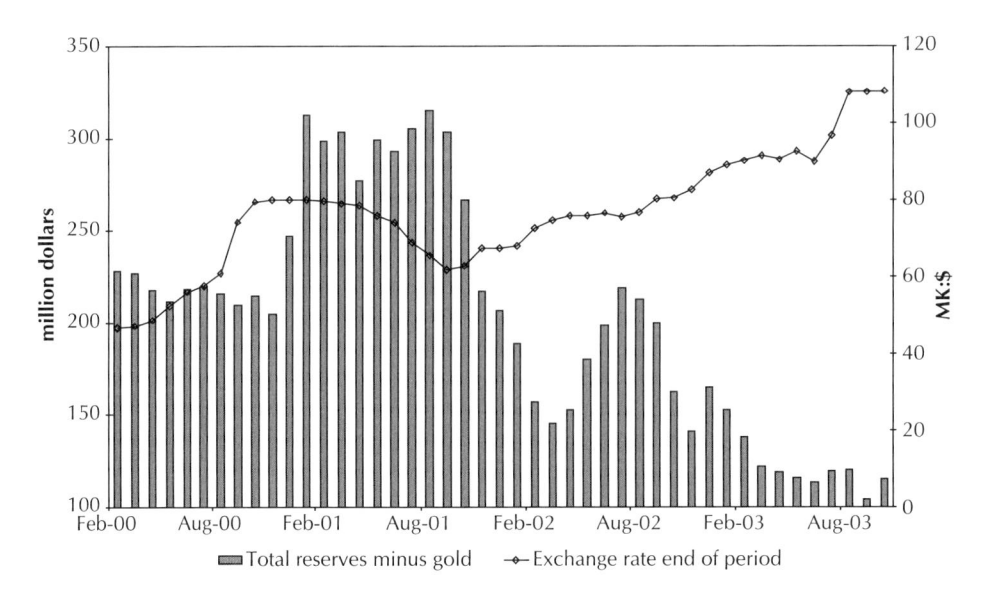

Fig. 15.1. Reserves and exchange rate, 2000–2003. (IMF, International Financial Statistics.)

tained. By August 2002, the landed price of maize was MK25 per kg and rising, so continuing to sell at MK17 per kg in 2002/03 clearly implied a consumer price subsidy. The IMF backed the subsidy because of the crisis situation. In the event, the MK17 per kg price was maintained by ADMARC until June 2003, but it was only able to sell 37,000 t of maize at this price because of large-scale food aid and private sector imports.[4] In an attempt to recoup some of the losses sustained by public finances, most of the remainder was sold at MK10 per kg in 2003/04.

The government miscalculated the scale of the imports and the level of consumer price subsidy required in 2002/03, partly because of weak management capacity and partly because of poor communications with donors, who moved from under-reaction in 2001/02 to a major food relief operation in 2002/03. However, it got one thing right: if maize was to be imported, it would have to be sold at a subsidized price. A non-subsidized price would have been over MK30 per kg, and the experience of 2001/02 had shown that these price levels lead to extreme food insecurity during the hungry period. However, importing maize for sale at a subsidized price inevitably implies fiscal losses, as the government cannot recover the full cost of the imports from sales revenue.

15.3 The Problem of the Exchange Rate

So what had changed? The key factor was the exchange rate. The nominal value of the Malawi Kwacha declined by 66% between end-August 2001 and end-August 2003. This meant that, even if dollar prices of maize in nearby markets and transport costs remained the same, rural incomes would have had to increase by 66% over a 2-year period to allow consumers comparable levels of access to the imported maize without price subsidies. Although there are no reliable data available on rural incomes, this clearly did not happen. In other words, rural consumer *demand* for imported maize at market prices was substantially reduced by exchange rate depreciation, although *need* remained strong.[5]

We conclude that maize imports are no longer a solution to hunger crises in Malawi because:

- rural consumers cannot afford imported maize without large price subsidies;
- the government cannot afford to import maize and subsidize sales; and
- if the government does resort to imports and consumer price subsidies, this seriously undermines macro-economic stability, including the exchange rate.

Looking to the future, would it be possible to stabilize the exchange rate so that local incomes could catch up and imported maize would become affordable without consumer price subsidies? It is hard to envisage this at present because (i) the RBM is committed to allowing the market to determine the value of the Kwacha; and (ii) even if it wanted to slow down exchange rate depreciation, it would be highly unlikely to succeed due to the country's weak foreign reserves and balance of payments position.

Can the private sector import maize and sell it at market prices? We can expect some cross-border flows from neighbouring countries to continue, but rapid depreciation of the Malawi Kwacha is bad for maize imports from Tanzania, Zambia and Mozambique. As long as import parity prices remain high in local currency terms, the market in Malawi will be restricted to relatively wealthy consumers, so cross-border imports will not solve the country's food security problems. The rural poor will remain unable to buy imported maize. One solution might be to artificially raise poor consumers' purchasing power by providing cash transfers during the hungry period, but the benefits of this approach might be eroded by consumer price inflation (see Section 15.6).

15.4 Food Aid

Can the country rely on large-scale food aid instead of commercial imports? Between June 2002 and April 2003, the World Food Programme (WFP) imported 169,000 t of maize. Together with 31,000 t purchased locally, making a total of 200,000 t, this maize was distributed via non-governmental organizations (NGOs) as part of a free food aid package for the 'hungry period'.

This type of intervention reaches those who need it and also helps to keep maize prices low by reducing demand-side pressure on the market (Levy, 2003). Moreover, food aid does not damage public finances or undermine the exchange rate. However, the cost is high, albeit a cost which is paid by taxpayers in developed countries. The cost of importing and distributing food[6] through the WFP and NGO operation in 2002/03 was around US$500/t (including overheads). For this reason, it is not realistic to deal with *chronic* food insecurity this way.

However, food aid is, at present, the best option for Malawi in cases of severe harvest failure.[7] Early warning signals (see Chapter 6) should be monitored carefully to allow as much time as possible to organize aid inflows and distribution. A critical food shortage should be predictable by harvest time in any year, allowing 6 months' planning before the hungry period.

15.5 Producing More Maize

If formal commercial maize imports and food aid are expensive solutions for Malawi's chronic food shortages, is domestic production a better option? This section examines the alternatives open to policy makers.

The conditions faced by smallholder farmers nowadays are very different from those of 10 years ago, at the end of the Kamuzu Banda era. Agricultural liberalization has produced huge changes. Farmers no longer enjoy input subsidies or have access to government credit schemes, and the exchange rate has depreciated sharply, making imported fertilizer much more expensive (see Introduction). At the same time, repeated food crises have persuaded decision makers of the need to keep maize prices relatively low to avoid hunger (see Chapter 8). As a result, maize is no longer a profitable crop for smallholders. Moreover, since the end of maize procurement at fixed prices by ADMARC in 1994, maize prices have been volatile, so there is considerable risk involved in growing maize commercially.

Assuming that it would be incompatible with the government's commitment to agricultural liberalization to return to a system of maize procurement at fixed prices,[8] there are two options for solving the maize production problem:

1. A general subsidy on the price of fertilizer; or
2. Starter Pack – a targeted production (fertilizer and seed) subsidy.

Both of these options are designed to boost national maize production by tackling the problem of expensive inputs. The first aims to boost household maize self-sufficiency and to make maize production profitable once more, leading to increased supplies of maize in the market. Whether it would achieve this in the post-agricultural liberalization context is hard to predict. The second option is a tried and tested one. It is consistent with a strategy of ensuring food security for poor smallholder farmers by boosting household production, reducing the need to purchase food in the hungry period and keeping prices low. The options are discussed below, and their costs and benefits summarized and compared with the other alternatives (imports, food aid and safety nets) in Table 15.2.

15.5.1 A general fertilizer price subsidy

Bingu wa Mutharika, who was elected president of Malawi in May 2004, has promised to reintroduce a general fertilizer price subsidy. Theoretically, this could have two benefits: first, to enable medium-sized smallholder farmers to overcome their input constraint and produce more maize for home consumption (depending on the level of the subsidy, some might even become self-sufficient); and second, to enable smallholder farmers with larger cultivated land areas to produce maize for sale at a profit. However, it has to operate within a context of relatively low maize prices if it is to avoid hurting smallholders with small land areas, who will remain deficit producers even if they have access to fertilizer.

We can make a rough calculation of who would benefit from this intervention: the first group would be farmers cultivating between 2 and 3 acres in the main season (some 25% to 30% of all smallholders[9]); the second group would be smallholder farmers with cultivated areas of 3 acres or more (around 30% nationally, but less than 15% in the southern region), as well as the estate sector comprising over 30,000 larger farms. A third group consisting of the poorest smallholders with less than 2 acres (over 40% of smallholders

Table 15.2. Comparing the costs and benefits of national food security interventions.

	Cost per year	Food security benefits	Limitations/problems	Uncertainty	Risks
1. Commercial maize imports sold at subsidized prices	Some US$70–US$100 million for 300,000 t. Cost per tonne US$245–319. Possibility of small proportion of cost recovery through sales.	Increases availability of maize and keeps prices low.	• High fiscal cost • Erodes foreign reserves and exchange rate • Undermines private sector and growth	• Difficult to predict amounts required and level of subsidy needed • Cost of imports vary • May not recoup costs	LOW
2. Food aid (approximately 80% maize)	Around US$100 million for 200,000 t (less required because targeted to poor in rural areas). Cost per tonne approximately US$500.	Delivered direct to those in need; reduces demand pressure and keeps food prices low.	• High cost to donors • Unlikely to be available year in, year out (only available in cases of severe harvest failure)	• There could be a delayed donor reaction to warning signals, as in 2001/02	LOW
3. General fertilizer price subsidy	Starts at US$20 million for a 70% subsidy on urea and basal, but *costs could escalate* depending on demand.	*Should* (if used for maize) help medium-sized farmers to produce more maize for home consumption and larger farmers to produce for sale at a profit, increasing marketed maize supply.	• Does not benefit the poorest smallholders • Impact in context of liberalized agriculture is unknown: it may do very little to boost maize production or sales.	• Demand for subsidized fertilizer cannot be forecast with any degree of accuracy	HIGH • Escalating costs could force government to withdraw subsidy • Fertilizer may be used on tobacco instead of maize • Vulnerable to occasional harvest failure

Financing and Macro-economic Impact

4. Starter Pack	Approx. US$20 million (fixed cost) for 2.8 million Starter Packs. Cost per additional tonne of output: approximately US$57.	*Proven* capacity to boost maize production; achieves food security for the poor by reducing demand pressure and keeping food prices low.	• Pack is too maize-focused at present (diversification needed)	• Negligible	LOW • Vulnerable to occasional harvest failure
5. Safety nets: DWTs (cash or food)	Around US$107 million to provide MK2000 per month for 30% of households for 5 months of hungry period	Received by the most needy, assuming targeting is successful.	• Difficult to ensure that most needy are targeted • Major capacity required to administer on seasonal basis (5 months only)	• There will be year-to-year variations in 'right number' of beneficiaries (30% is only a rough guide)	MEDIUM • Injections of cash or efforts to source food could cause *inflation* if supply shortages exist

Note 1: The scale of each intervention is roughly that which would be needed to avoid extreme food insecurity in the 'hungry period'.
Note 2: Only one example of a safety net has been included: DWTs. A conservative estimate is that it would need to reach 30% of households; this assumes successful targeting.

nationally, rising to 60% in the southern region) would be unlikely to benefit at all, as they could not afford to buy fertilizer even with a subsidy in place.

The main problem, however, is the level of subsidy required. With the cost of fertilizer for the 2004/05 season around US$20 per 50 kg bag, the 10% to 20% subsidy levels of the Banda era would have almost no impact on either production for home consumption or production for sale – because with this level of subsidy most farmers would still not be able to afford the fertilizer, and commercial production would not be profitable. Although there is insufficient data to accurately estimate the size of the subsidy that would be required to make a serious impact, a rough calculation suggests that it would need to be in the order of 70%. However, it is difficult to predict how much additional maize might be produced at this, or any other, level of subsidy.

Another serious problem with a general fertilizer price subsidy is the unpredictability of demand for fertilizer. The IFDC reported that 39,773 t of basal and 46,199 t of urea were sold in Malawi in 2002 (the last year for which data are available). If similar amounts had been sold at 2003 prices with a 70% subsidy, the cost to the government would have been MK2.2 billion (US$20 million). However, if demand were to increase substantially because of the subsidy, quantities sold could escalate and the fiscal cost could be much higher. Efforts to calculate demand for fertilizer at different levels of subsidy will be unreliable, as the data does not exist that would allow economists to make accurate forecasts. This means that the MoF could be faced with a much larger bill than predicted.[10]

A modification of this approach would be to combine a lower level of general fertilizer price subsidy with efforts to keep the price of maize relatively high and stable (see Chapter 19). This would cost less in terms of the fertilizer subsidy (although the cost would still be hard to predict), but there would be additional costs of:

1. Intervention in the maize markets to stabilize prices, e.g. through the SGR; and
2. Providing safety nets for the 'casualties' of the approach: poor farmers who would be unable to afford the fertilizer to produce enough food on their own farms *and* who would face relatively high maize prices during the hungry period.

The additional cost of intervention in the maize markets to stabilize prices is hard to calculate, and therefore to budget for. The handling costs for 100,000 t of maize stored in the SGR are roughly US$1.2 million per annum, while the possible losses from large-scale buying and selling operations are impossible to quantify and open to abuse. On the other hand, the cost of providing safety nets for the poorest, most food insecure farmers is predictable and would be high (see Section 15.6).

However, much of the above discussion would probably be academic, as it is likely that most of the subsidized fertilizer would end up being used for tobacco farming rather than for maize (even if the subsidy were restricted to basal and urea, which is appropriate for growing maize). This would make demand, the cost of the intervention and results in terms of increased maize production and food security even harder to predict.

We conclude that on whatever criteria the idea of a general fertilizer price subsidy for Malawi is judged – food security outcome, fiscal cost (and therefore impact on other macro-economic variables), or uncertainty and risk – it is clearly a bad option.

15.5.2 Starter Pack – a targeted production subsidy

By contrast, Starter Pack has proven benefits in terms of food security and social protection (see Chapter 8). It is a broadly targeted subsidy in that it goes only to smallholders, not to estates. The tiny pack of inputs, enough for around 0.1 ha of maize, benefits virtually all smallholder households, including the poorest and most vulnerable in the remotest areas of the country. The scaling down of Starter Pack in the 2000/01 season and the late delivery of the packs to beneficiaries were major contributing factors to the 2001/02 food crisis (see Introduction). This constituted a failure by government and donors to maintain a commitment to tackling food insecurity from the production side.

Universal Starter Pack (for 2.8 million households) has a total cost of some US$20 million and a cost per tonne of additional maize produced of around US$57, which compares favourably with the cost per tonne of imported maize and food aid (see Table 15.2). Costs are clearly accounted for by the Logistics Unit (see Chapter 3). All of the items of expenditure required for the programme in any given fiscal year are easy to cost at the time of drawing up the budget in May/June, since the quantities involved are fixed once the number of packs to be distributed is agreed upon. If the programme were to be incorporated into the Medium Term Expenditure Framework (MTEF), annual budgeting would become even easier.

Starter Pack is highly efficient. Around 80% of the cost of the programme goes directly to the beneficiaries (overhead costs are around 20%). Moreover, the implications for public expenditure management are simple: the Logistics Unit has 6 years of experience, and most of the initial problems have been overcome (see Chapter 3).

As the programme has been the subject of in-depth evaluations, there is little that we do not know about it, so unpredictability is not an issue. There is little risk of future Starter Pack programmes failing to meet food security objectives except in the case of severely adverse weather conditions (e.g. serious, widespread drought) causing harvest failure.

15.6 Safety Nets

Safety nets such as Direct Welfare Transfers (DWTs), Public Works Programmes (PWPs) and child feeding programmes are seen by some members of the donor community as alternatives for achieving food security in

Malawi. This section argues that, although they can play important roles in helping the most vulnerable, developing rural infrastructure, promoting private sector development and enhancing rural incomes, they represent an inefficient and in some cases ineffective way of solving the problem of chronic national food insecurity.[11]

Why are safety nets an *inefficient* way of solving chronic food insecurity? We will examine a simple type of safety net: DWTs. A pilot programme in Dedza in 2001/02 found that it was possible to target the 'work-constrained poor' to receive DWTs of cash or food (Levy *et al.*, 2002), but the cost of scaling up such a programme to reach enough households to have an impact on food insecurity was high. A reasonable assumption about the scale required for this purpose would be around 30% of rural households, or approximately 0.8 million.[12] If we assume that a DWT intervention should provide these households with MK2000 per month[13] for 5 months of the year (the hungry period), this would cost MK10,000 (US$95) per household per year, plus overheads of 40%, as in the Dedza pilot, making a total of US$107 million. Management of DWTs on such a scale would be extremely complicated, and there are serious uncertainties and risks (see Table 15.2), not least of which is that they are likely to lead to inflation if food is in short supply.

Why are safety nets sometimes *ineffective* as food security interventions? The most widespread form of safety net intervention in Malawi is the PWP. These work reasonably well on a small scale, but they cannot be scaled up to the sort of levels that would make an impact on the country's chronic food insecurity. To have a national food security impact, PWPs would need to be implemented in most of the country's 30,000 villages every year. Providing employment and food, cash or agricultural inputs in Village A one year and moving on to Village B the following year would not work, as the food security benefits are not likely to be sustained after the PWP is withdrawn.

Safety nets can, however, complement large-scale food security interventions by supporting a small percentage of the poorest and most vulnerable households. Where they provide the beneficiaries with food, a related issue is how this should be sourced. The NFRA, which manages Malawi's SGR, is currently proposing that reserves be kept at a low level – around 30,000 t of maize – with the emphasis on supplying maize to the WFP and NGOs for use in safety net interventions, rather than on maintaining 'buffer stocks' or engaging in price stabilization. To keep costs down, purchases should be mainly from local estates (under contract farming arrangements) rather than from abroad.

15.7 Conclusion

Malawi's food security policy options are now much more limited than 10 years ago because of a sharp deterioration in the macro-economic environment, in particular the exchange rate. Rapid exchange rate depreciation (raising the price of fertilizer in local currency terms) without compensating increases in rural incomes has undermined local maize production. It has also reduced the possibility of using imported maize to bridge the food gap without consumer price subsidies, which the government cannot afford.

The Malawi Government's response to the 2001/02 and 2002/03 food shortages had extremely serious, negative macro-economic implications. Some of the adverse impact could have been avoided by better management and by improved government–donor coordination. However, most of it was inevitable because formal maize imports and subsidized consumer prices were used to bridge the food gap, and these cost money which the government did not have.

Formal maize imports were only needed in 2001–2003 because of a failure to tackle food insecurity from the production side. This chapter has argued that a key lesson for the future is that importing food from South Africa is a very expensive way for Malawi to deal with chronic food insecurity. A much less costly approach is to prevent food crises before they happen. This means providing some sort of government subsidy to domestic production. Our view is that the best option is a targeted production subsidy (Starter Pack) rather than a general fertilizer subsidy. Starter Pack – costing some US$20 million annually for a universal programme reaching 2.8 million beneficiaries – compares extremely well on cost with alternative food crisis prevention measures such as general fertilizer price subsidies, as well as with relief interventions such as subsidized commercial food imports and food aid. Safety nets are not an alternative for achieving national food security. We conclude that Starter Pack should be financed as part of Malawi's MTEF.

Notes

[1] There have been clear signs of increasing production of the country's second staple food crop – sweet potatoes – in recent years, and cassava is well established (see Chapter 12). But total volumes produced are difficult to estimate, and over-optimism can be dangerous. In 2001, a European Commission (EC)-sponsored crop assessment survey report (Sichinga et al., 2001) massively overestimated cassava production, leading the EC to underestimate the size of the food gap in 2001/02 and to fail to take action in time to prevent the food crisis.
[2] In FY2001/02, there were reported to have been cuts in other areas of expenditure, even Pro-poor Expenditures (PPEs). However, much of this was attributable to factors other than maize imports. From 2002/03 onwards, PPEs (health, education, extension, etc) have been protected from cuts, following objections by the IMF to what happened in 2001/02.
[3] The US$75 million translated into MK6.1 billion at the time of purchase, but with interest and marketing charges the total bill came to MK7.4 billion.
[4] In 2002/03, private importers reacted to the food shortages. Informal, private imports from neighbouring Mozambique, Zambia and Tanzania reached an estimated 246,000 t, up from 155,000 t in 2001/02, according to Whiteside et al. (2003).
[5] Demand refers to the combination of need and the capacity to realise it (purchasing power).
[6] Some 80% of the food basket was maize.
[7] Other options are to build up 'buffer stocks' in the country's SGR or to take out some form of crop failure insurance or weather insurance (Devereux, 2003a,b). However, donors in Malawi feel that buffer stocks are not an option because they are expensive and open to abuse, while the proposed insurance mechanisms are only at the discussion and piloting stages.

[8] A return to the old ADMARC system of fixed producer prices would also imply a major financial burden (and risk) for the government, but this is beyond the scope of this chapter.
[9] The proportions are calculated using the 2003 TIP survey database (List A beneficiaries).
[10] In the 2004/05 season, the government piloted a fertilizer price subsidy scheme in which 500,000 registered beneficiaries were given vouchers entitling them to a subsidy. In the author's opinion, the use of vouchers based on a beneficiary register does not amount to 'targeting' the benefit, as some government officials claim, but it should avoid the problem of unpredictability of cost for the MoF.
[11] A detailed discussion of safety net interventions in Malawi is beyond the scope of this chapter, which merely considers their potential in relation to the chronic food insecurity problem.
[12] Numbers of beneficiaries are difficult to estimate accurately, as data on numbers of rural households and poverty levels are disputed. The National Safety Nets Strategy (NSNS) works on the assumption of around 25–30% extreme poverty, which is roughly consistent with the findings of recent TIP evaluations on levels of extreme food insecurity in rural areas. The NSNS does not propose to provide DWTs to 25–30% of households, but rather aims to provide them with a mixture of different types of safety net. However, it is reasonable to assume that a mixture of safety nets (DWTs, PWPs and child feeding) sufficient to protect 0.8 million households from hunger would be at least as expensive as providing them with DWTs.
[13] This figure is recommended in a recent study on wage rates for PWPs (Chirwa et al., 2004).

References

Chirwa, E., McCord, A., Mvula, P. and Pinder, C. (2004) Study to inform the selection of an appropriate wage rate for public works programmes in Malawi. Report for the National Safety Net Unit (GoM), MASAF and CARE. Lilongwe, Malawi. Overseas Development Institute, London.

Devereux, S. (2003a) Malawi's food crisis of 2001/02: implications for food security policy. Work in progress briefing for Action Aid. (Unpublished.)

Devereux, S. (2003b) Policy options for increasing the contribution of social protection to food security. ODI Forum on Food Security in Southern Africa. Overseas Development Institute, London.

Levy, S. (2003) Starter Packs and hunger crises: a briefing for policymakers on food security in Malawi. (Unpublished, available on CD in Back Cover Insert.)

Levy, S. (2004) The Malawi government's response to the 2001/02 and 2002/03 food crises: maize imports, financing and macroeconomic impact. (Unpublished.)

Levy, S., Nyasulu, G. and Kuyeli, J. with Barahona, C. and Garlick, C. (2002) Dedza Safety Nets Pilot Project: learning lessons about direct welfare transfers for Malawi's National Safety Nets Strategy, final report. (Unpublished, available on CD in Back Cover Insert.)

Levy, S. with Barahona, C. and Chinsinga, B. (2004) Food security, social protection, growth and poverty reduction synergies: the Starter Pack programme in Malawi. *Natural Resource Perspectives* No. 95. Overseas Development Institute, London.

Sichinga, K., Salifu, D. and Malithano, D. (2001) EC food security and food aid programme crop assessment report 2000/2001, Vol. 2, final report. Agricultural Policy Research Unit, Bunda College, University of Malawi, Lilongwe, Malawi.

Stevens, C., Devereux, S. and Kennan, J. (2002) The Malawi famine of 2002: more questions than answers. Institute of Development Studies, Brighton, UK.

Whiteside, M. with Chuzo, P., Maro, M., Saiti, D. and Schouten, M.-J. (2003) Enhancing the role of informal maize imports in Malawi food security. Report for DFID-Malawi. (Unpublished.)

Whitworth, A. (2004) Malawi's fiscal crisis: a donor perspective. DFID, Lilongwe, Malawi.

IV By Special Invitation

16 Poverty, AIDS and Food Crisis

Anne C. Conroy

16.1 Introduction

This chapter explores the links between poverty, AIDS and food crisis in Malawi. The first part of the chapter describes the extent of poverty in the country, the poor nutritional indicators and the apparent worsening of poverty in recent years. The second part outlines the dimensions of the AIDS pandemic. Despite the fact that HIV prevalence rates are stabilizing at 12–17% of the adult population, the pandemic is having a devastating impact on Malawi's people and economy. The AIDS pandemic also contributes significantly to food insecurity by undermining labour productivity and diverting resources away from agricultural investment into health care and funerals.

The 2001/02 food crisis led to an increase in high-risk sexual behaviour and the incidence of sexually transmitted infections, and to a deterioration in maternal health. This in turn fuelled the AIDS pandemic. Malawi is suffering from a vicious cycle, as food crises undermine economic growth and human development in ways that are not fully understood. A universal Starter Pack implemented consistently for several years could help to break this cycle and support the government's poverty-reduction strategy.

16.2 Poverty

According to Malawi's 1997/98 Integrated Household Survey (IHS), 60% of Malawian households are 'poor': they do not meet basic nutritional and other requirements; some 29% of households are 'ultra poor', meeting less than 60% of their basic needs (National Economic Council, 2000). The incidence of poverty is highest in rural areas: 61% of the rural population is poor compared with 51% of the urban population.[1]

The Gini coefficient[2] for Malawi as a whole is 0.401, while the rural Gini coefficient is 0.374. This means that Malawi does not have high income inequality compared with other countries in the region (see Table 16.1). Nevertheless, the richest 20% of the rural population in Malawi accounts for 44% of total consumption, while the poorest 20% accounts for only 7% of total consumption. In 1997/98, average daily per capita income was MK4.45 (US$0.14) and ultra-poor households had average daily per capita income of MK1.21 (National Economic Council, 2000).

Malawi's nutritional indicators are very poor, reflecting poverty and persistent food insecurity. At the national level, 62% of young boys[3] and 57% of young girls are stunted, reflecting chronic malnutrition, while 9% of young boys and girls are wasted, reflecting acute malnutrition. The figures are the same in rural areas (National Economic Council, 2000). Food intake is inadequate. The IHS found that 'poor' households consume only 66% of their recommended daily requirement (RDR) of calories, while 'ultra poor' households consume only 54% of their RDR (the figures are the same at national level and for rural areas). The diet is dominated by cereals, which account for almost 80% of calories consumed by poor households. The poorer the household, the more the diet is cereal-dominated. Income levels are staggeringly low in rural areas and expenditure on food dominates household expenditure. Poor households are very vulnerable to increases in the price of cereals, in particular the most widely consumed staple food: maize.

Available evidence suggests that poverty has worsened in Malawi over the past decade. The underlying causes of poverty include persistent food insecurity, declining agricultural productivity, poor literacy, limited options for off-farm or formal-sector employment, declining terms of trade for principle export crops, inflation (which erodes the incomes of the poor), very poor health indicators, a high disease burden, a devastating AIDS pandemic and increasing death rates. Poor households are most affected by AIDS and other

Table 16.1. Poverty and Gini coefficients for selected countries in the region.

Country (survey year)	% of total population under poverty line	% of rural population under poverty line	Gini coefficient (national)
Malawi (1997/98)	60	61	0.401
South Africa (1995)	–	–	0.593
Tanzania (1991*; 1993**)	39	41	0.382
Zambia (1996*; 1998**)	69	83	0.526
Zimbabwe (1990/91*; 1995**)	26	36	0.568
Mozambique (1996/97)	69	71	0.396

Note: Poverty lines are constructed separately in each country, and are not directly comparable.
* Year of poverty line calculation; ** Year of Gini coefficient calculation.
Source: Malawi – National Economic Council (2000); other countries - World Bank: World Development Indicators (2004).

infectious diseases as they have insufficient food, limited peace of mind and little money to pay for medical care. The AIDS pandemic contributes to poverty by undermining economic growth and eroding gains from investment in human development. It causes untold human misery and affects all families in Malawi.

16.3 HIV/AIDS

The analysis of sentinel surveillance data collected in 2003 (National AIDS Commission, 2003, 2004) indicates that HIV prevalence in Malawi is in the range of 12–17% of the adult population.[4] Prevalence rates are stabilizing, but this does not mean that the problem has gone away. There are over 100,000 new infections per year, and there were an estimated 87,000 AIDS deaths in 2003. Prevalence is higher in urban areas (19–28%) than in rural areas (10–15%). HIV prevalence is also twice as high in the more densely populated southern region (around 20%) than in the central and northern regions (around 10%).

The National AIDS Commission (2004) estimates that some 88% of infections are due to heterosexual intercourse, while 10% are due to perinatal transmission. Transmission rates due to other means (unsafe blood products, inadequate observation of universal precautions and homosexual intercourse) are unknown, but thought to be negligible.

Despite almost universal knowledge about how the virus is transmitted and how to reduce the probability of transmission, there has been very slow progress in terms of changing high-risk sexual behaviour. Young people are especially vulnerable to infection, and gender inequality, lack of social cohesion, poverty and inequality all drive the pandemic (Conroy et al., forthcoming).

The implications of the pandemic are devastating for Malawi. According to the National AIDS Commission (2003, 2004):

- Over 840,000 children under the age of 18 are orphans, with 45% due to HIV/AIDS.
- Some 70,000 children (0–14 years) are infected with HIV. There were an estimated 25,840 new infections, 20,600 new AIDS cases and 20,410 deaths due to AIDS among children in 2003.
- The death rate for adults (15–49 years) has tripled since 1990.
- The number of cases of tuberculosis (TB) is three times higher than it would be without HIV/AIDS.
- 170,000 people are in presently in need of anti-retroviral therapy, and an even larger number of people need voluntary counselling and testing to learn their HIV status.
- About 500,000 pregnant women need good antenatal care, including counselling and testing.
- About 80,000 mothers need anti-retroviral therapy to prevent vertical transmission of the virus.

The impact of the epidemic will get worse even with a stable prevalence rate. By 2010, the number of annual AIDS deaths is projected to reach 96,000, while the number of people needing anti-retroviral therapy is expected to reach 190,000 (National AIDS Commission, 2003, 2004).

16.4 How AIDS Contributes to Food Crisis

Smallholder agriculture in Malawi depends critically on labour. AIDS affects agriculture because the farm labour force is made up principally of sexually active adult men and women; AIDS exacerbates an already serious labour constraint, reducing labour availability and productivity. Women face the 'double burden of care' as they are more likely to be infected with HIV/AIDS and are also responsible for most of the agricultural labour and for caring for the chronically sick. AIDS also has an adverse impact on agriculture by diverting capital from investment, depleting assets and diverting income to pay for the cost of health care and funerals. Communities are devastated emotionally and practically by the amount of time that is spent at funerals. In Malawian culture, it is very important for community members to attend funerals, and fields are deserted as the whole community pays its respects to the dead. AIDS also erodes knowledge on how to adapt to new circumstances and manage communal resources, as well as knowledge about farm management, soils and environmental conservation (Richards, 1999). Finally, AIDS erodes the capacity of institutions that support agriculture in the area of planning, extension, research and market information.

De Waal and Whiteside (2003) argue that AIDS progressively undermines the resilience of smallholder agricultural systems in southern Africa. They suggest that AIDS is fuelling a 'new variant famine'. In normal famines, deaths are concentrated in the young and the very old. By contrast, AIDS leads to famine deaths among economically productive adults. AIDS therefore triggers an economic crisis and impairs the ability of institutions to provide public services. In addition, AIDS exacerbates household labour shortages and erodes skills due to adult mortality and morbidity. It also leads to an increased burden of care as a significant minority of the population are chronically sick. The AIDS pandemic destroys resilience and traditional coping mechanisms. New coping mechanisms, including the sale of productive assets and commercial sex work, undermine prospects for recovery and fuel the transmission of HIV. Finally, there is the vicious interaction between malnutrition and HIV/AIDS, as undernourished individuals are more susceptible to infection. Nutritional status is also an important factor determining the probability of mother to child infection (de Waal and Whiteside, 2003).

Research in central Malawi confirms the devastating impact of AIDS on livelihoods. AIDS triggers a sequence of impacts on the rural economy, including leaving land fallow, delaying agricultural operations and changing the crop mix to include less labour-intensive and valuable crops. All these reduce the value of agricultural production, undermine food security and exacerbate poverty. The negative impact of AIDS on rural livelihoods is greatest if the onset of sickness coincides with the start of the agricultural season, as labour and capital are diverted to care for the sick individual (Shah *et al.*, 2002). The household can no longer do *ganyu* and may have to sell assets or resort to distress sales of crops (selling for very low prices), forcing it deeper into poverty. The impact is also exacerbated when both parents get sick and die in quick succession, which may lead to the dissolution of the household.

Children face grief, loneliness and trauma as their parents become sick and die. As households dissolve, the orphans often face insecurity and stress in new households. Despite the generosity of Malawian households and their willingness to care for orphans, the extended family system is failing to cope with the sheer numbers of adult deaths. Grandmothers may be caring for up to ten grandchildren, and there are a growing number of orphan-headed households. Orphans are very vulnerable; they may be forced to drop out of school and are also vulnerable to abuse.

16.4.1 The 2001/02 crisis

Food security is extremely precarious in Malawi. The majority of rural households are food insecure for at least 3 months of the year even in 'normal years'. In March 2002, the Vice President of Malawi appealed to the donor community: 'Coping mechanisms including casual labour and distress selling of household assets have been stretched and eroded, while people are dying of hunger related diseases' (Malewezi, 2002).

The AIDS pandemic did not cause the 2001/02 food crisis. The underlying causes of food insecurity in Malawi include the severe land constraint, declining soil fertility, limited options for off-farm employment and the declining profitability of the country's principal export crops (tobacco, sugar and tea). More immediate factors contributing to the crisis were the scaling down of Starter Pack and the Agricultural Productivity Investment Programme (APIP, see Chapter 17) and poor management of the Strategic Grain Reserve in 2001. Weak agricultural information systems, inadequate response mechanisms and the lack of a comprehensive programme for social protection also contributed significantly. However, the AIDS pandemic had eroded livelihoods and made households more vulnerable and less able to cope with shocks like the poor 2001 maize harvest and the unprecedented rise in food prices in the 2001/02 hungry period.

16.5 How Food Crisis Fuels the AIDS Pandemic

The 2001/02 food crisis fuelled the AIDS pandemic directly through its impact on high-risk sexual behaviour and indirectly through its impact on government finances.

During the first 6 months of the food crisis, there was a significant increase in high-risk sexual behaviour. According to the United Nations

Humanitarian Response (UNHR) Malawi team (2002), adult men increased mobility as they sought opportunities for off-farm employment. This increased mobility led to an increase in the number of sexual partners. Women increasingly turned to transactional sex in order to get food to feed the family, and young women were often pressurized into having transactional sex or forced into prostitution or early marriage. Focus group discussions also revealed a significant increase in gender-based violence, especially in polygamous households[5] (UNHR Malawi, 2002). Such factors, combined with the breakdown in social cohesion in the initial stages of a food crisis, all fuel HIV transmission.

The increase in high-risk sexual behaviour associated with the food crisis led to a significant increase in sexually transmitted disease, teenage pregnancy and complications of abortion in areas worst affected by the crisis (UNHR Malawi, 2002):

1. The number of people reporting to hospitals with sexually transmitted infections increased by 31%.
2. The number of spontaneous abortions increased by 62%.
3. The number of cases of complications of abortion increased by 114%.
4. The number of cases of anaemia increased by 96%.
5. The number of cases of haemorrhage increased by 153%.

This, in turn, meant that there would be an increase in maternal mortality rates, which were already rising before the crisis: the maternal mortality rate increased from 620 per 100,000 live births in the 1992 Malawi Demographic and Health Survey (MDHS) to 1120 per 100,000 live births in the 2000 MDHS (National Statistical Office, 2001).

Thus, the 2001/02 food crisis increased vulnerability to HIV/AIDS by increasing pressures for unsafe sexual practices, increasing the incidence of teenage and high-risk pregnancies and increasing gender-based violence. Any deterioration in nutritional status also increases susceptibility to diseases including AIDS (it weakens the immune systems of HIV-positive individuals). The UNHR Malawi team (2002) reported that:

> The WHO-led health assessment carried out in September 2002 showed a 36% increase in new TB cases (a good proxy indicator for HIV since co-infection exists in over 80% of cases). Other relevant information was a crude mortality rate (CMR) of 1.96 per 10,000 people per day (> 1 is considered a humanitarian emergency), and an under-5 CMR of 3.8 (> 2 signifies a humanitarian crisis).

The food crisis also reduced incentives for health-seeking behaviour because many people were too weak to walk to the nearest health centre (UNHR Malawi, 2002).

The UNHR team's survey revealed that the health sector was not in a position to respond adequately to the impact of the crisis on the sexual and reproductive health needs of the general population and people living with HIV/AIDS. The report concluded that:

without prompt attention to sexual and reproductive health needs within the context of the humanitarian response, the impact of the food crisis will be extended well beyond the duration of the food shortage by increasing the burden of ill health in the general population, particularly in the area of maternal and newborn health, HIV transmission and AIDS deaths (UNHR Malawi, 2002).

The food crisis also exacerbated the macro-economic crisis. Government imported maize at a cost of US$33 million in 2001/02 and US$75 million in 2002/03, which it financed by printing money and by huge increases in domestic debt (see Chapter 15). In 2001/02, this meant that there was less money available for health spending – although from 2002/03, health budget allocations were protected as a 'Pro-poor Expenditure'. The food crisis also exacerbated the capacity constraint in government by diverting time and resources away from other activities. It is difficult to underestimate the management, time and coordination challenges posed by the crisis. The government, donors, non-governmental organizations (NGOs) and civil society collaborated well and responded effectively, but all energies were focused on responding to the crisis rather than on economic growth and poverty alleviation.

Finally, the food crisis depressed overall economic growth rates and pushed households deeper into poverty. There were serious human and educational costs, as the crisis undermined child development through increases in acute malnutrition, impaired learning capacity and dropping out of school early. The food crisis also undermined the health status of the population, which will have a long-term impact on economic growth.

16.6 The Role of Starter Pack

Starter Pack was designed in response to a request by the Cabinet Committee on the Economy, which was concerned about avoiding a food crisis and ensuring that the government did not have to use scarce resources to import maize (see Chapter 1). The programme was highly successful in delivering national-level food security by increasing maize production and keeping maize prices relatively low, allowing access to maize for poor households in the hungry period. With a food secure population, the government could focus energy and resources on the other key priorities for poverty reduction set out in its Poverty Reduction Strategy Paper (Malawi Government, 2002).

At a household level, Starter Pack does not achieve food security for all. However, it does allow the poorest households with the greatest labour constraint – in particular those affected by HIV/AIDS – to use their labour productively and to produce more food than would be the case in the absence of the programme. By increasing food production by better-off households as well as poorer ones, it also helps to reinforce traditional safety nets within communities (see Chapter 8).

16.6.1 HIV/AIDS messages and the TIP evaluations

From the 2000/01 season, the Targeted Inputs Programme (TIP) has included messages on HIV/AIDS prevention, on the basis that AIDS is devastating rural communities and therefore any programme that has wide coverage should include a component to prevent transmission of HIV. In 2000/01, the Agricultural Communication Branch of the Ministry of Agriculture (MoA) worked with the National AIDS Commission to develop leaflets designed to coincide with World AIDS Day.

The AIDS component of the TIP has been evaluated comprehensively (Dzimadzi *et al.*, 2001; Chinsinga *et al.*, 2002). The studies shed light on the need to focus messages on HIV prevention and to make them clearer and more culturally acceptable to the target audience, especially to take account of inter-generational and cultural issues. The evaluation also highlighted the misconceptions that still persist about HIV transmission in rural Malawi and provided data on rural households' sources of information on HIV/AIDS. The findings indicated the importance of intensifying AIDS communication messages on the radio and identified gaps in knowledge on issues surrounding HIV/AIDS, especially among young people.

There is an urgent need to mainstream messages on HIV/AIDS prevention in all agricultural and rural development programmes. The TIP evaluation database contains information on many aspects of poverty, AIDS and food insecurity, which could be used constructively to inform future policy. However, some stakeholders take ideological positions for or against Starter Pack and, as a result, do not examine the wealth of data and analytical insights from the evaluation work, which could be used to improve the focus of future programmes for poverty reduction and HIV/AIDS prevention.

16.7 Conclusion

There is an emerging literature on how the AIDS pandemic fuelled the recent food crisis in southern Africa, but less understanding of how the food crisis will fuel the AIDS crisis – arguably in more direct and serious ways. The clear policy implication is that it is better to avoid a food crisis than to be forced to respond to one. Food crises undermine economic growth and human development in ways that are not fully understood.

A universal Starter Pack does not guarantee food security at the household level, but it does reduce the probability of severe national food crises, which undermine economic growth and fuel the AIDS pandemic. It also contributes to food security within households and reinforces community-level safety nets. If the programme were to be run consistently for several years, within a medium-term framework, this would allow the government to release resources and energies for poverty-reduction programmes. Without a powerful food security programme like Starter Pack in place, the Malawi Government will have less chance of meeting its goals under the Poverty Reduction Strategy.

Notes

[1] The information from the IHS presented here comes from the 6586 household data set.
[2] Gini coefficients measure the degree of equality or inequality in income distribution, expenditure and consumption. A coefficient of zero means that all households have the same level of resources (perfect equity). A coefficient of one represents the extreme of inequality, where all resources are concentrated in one household. Gini coefficients in Malawi's 1997/98 IHS are calculated using household consumption and expenditure.
[3] Age 6–59 months.
[4] Age 15–49 years.
[5] General levels of violence also increased as people resorted to theft to obtain food; when caught stealing, they were beaten viciously and in some instances killed.

References

Chinsinga, B., Dzimadzi, C., Magalasi, M. and Mpekansambo, L. (2002) 2001/02 TIP evaluation module 2: TIP messages: beneficiary selection and community targeting, agricultural extension and health (TB and HIV/AIDS). (Unpublished, available on CD in Back Cover Insert.)

Conroy, A., Malewezi, J., Sachs, J. and Whiteside, A. (forthcoming) *Africa's Perfect Storm: Poverty, HIV/AIDS and Hunger in the New Millennium in Malawi.*

de Waal, A. and Whiteside, A. (2003) 'New variant famine': AIDS and the food crisis in southern Africa. *The Lancet* 362, 1234–1237.

Dzimadzi, C., Chinsinga, B., Chaweza, R. and Kambewa, P. (2001) 2000/01 TIP evaluation module 3: agricultural communications. (Unpublished, available on CD in Back Cover Insert.)

Malawi Government (2002) Malawi poverty reduction strategy paper, Lilongwe, Malawi.

Malewezi, J. (2002) Malawi food crisis task force: appeal to the donor community, Lilongwe, Malawi. (Unpublished.)

National AIDS Commission (2003) Malawi national HIV and AIDS estimates 2003: technical report, Lilongwe, Malawi.

National AIDS Commission (2004) HIV and AIDS in Malawi: 2003 estimates and implications, Lilongwe, Malawi.

National Economic Council (2000) *Profile of Poverty in Malawi, 1998: Poverty Analysis of the Malawi Integrated Household Survey, 1997/98,* Lilongwe, Malawi.

National Statistical Office (2001) Malawi demographic and health survey 2000. Zomba, Malawi.

Richards, P. (1999) HIV/AIDS and smallholder agriculture: does technology have to change? In: Mutangadura, G., Jackson, H. and Mukurazita, D. (eds) *AIDS and African Smallholder Agriculture.* SAfAIDS, Harare.

Shah, M.K., Osborne, N., Mbilizi, T. and Vilili, G. (2002) *Impact of HIV/AIDS on Agricultural Productivity and Rural Livelihoods in the Central Region of Malawi.* CARE International, Lilongwe, Malawi.

United Nations Humanitarian Response Malawi (2002) Reproductive health and HIV/AIDS vulnerability assessment. UN Country Assessment Team (UNFPA, UNICEF, WHO, WFP, Ministry of Health and Population – Reproductive Health Unit), Lilongwe, Malawi.

17 Food Security Policies and Starter Pack: a Challenge for Donors?

JANE HARRIGAN

17.1 Introduction

Food security is a persistent preoccupation of all those involved in Malawi's development efforts. Government, donors and non-governmental organizations (NGOs) all acknowledge that a large proportion of Malawi's population remains chronically food insecure and that until this issue is addressed the broader process of development will be thwarted. However, within government and the donor community there are different views on how Malawi should best tackle the issue of chronic food insecurity. The aim of this chapter is to analyse these various strategies and to see how, if at all, the Starter Pack programme might fit into them. Although the government seems to have a fairly clear view on this, there is considerable division within the donor community. The analysis highlights the divisions and exposes the weaknesses in the donors' positions, which seem to be preventing a constructive debate on food security and the future of Starter Pack.

17.2 Food Security Policy Options

At the level of national aggregate food security, a country like Malawi faces two options: relying on domestic production (national food self-sufficiency) or relying on imports. These two sources of food are not mutually exclusive, in that part of the country's food requirements can be domestically produced and part imported. Over the past 10 to 15 years, 'Malawi has shifted from being a nationally self-sufficient producer of maize in non-drought years to being dependent on commercial food imports and foreign assistance to achieve a national food balance' (Øygard *et al.*, 2003). This poses a question to government and donors: Should this trend be allowed to continue, perhaps even be encouraged? Or should efforts be made to boost domestic food production?

If the import route is favoured, then polices need to be in place to ensure that foreign currency is available for food imports – either by promoting exportable cash crop production or via a process of industrialization. To the extent that foreign exchange earnings are inadequate for commercial food imports, food aid will be required.

If domestic production is the preferred route to national food security, a variety of policies (many of which are complementary) can be employed to boost domestic food crop production. These can include non-price policies such as technology transfer to improve yields, extension work, free inputs, infrastructure development, credit schemes and market development. Price policies, such as subsidized inputs and attractive output prices for food producers who market their crop (the latter might involve pan-territorial prices guaranteed by a state marketing board), can also be used. Decisions will also be needed about who to target and prioritize as food producers – subsistence producers, smallholders producing a marketable surplus or large-scale commercial farmers – and what food crops to prioritize.

Regardless of whether the emphasis is on imports or domestic production, vagaries in domestic production, exchange rates and import and transport costs make insurance policies desirable. These might consist of a strategic grain reserve and/or a financial reserve. Both may be needed if domestic production is unexpectedly low and import costs are unexpectedly high.

These are the main policy options for achieving national aggregate food security. However, such an achievement is not sufficient to ensure individual and household food security. A nation may possess adequate aggregate food supplies, whether domestically produced or imported, yet chronic food insecurity may persist if some people cannot grow enough food to eat or access food from the market. Policies are needed to ensure access to adequate food by all people at all times. This requires a livelihoods strategy combined with some form of social safety net for asset poor and vulnerable households.

A cross cutting issue in discussions of both national and individual/household food security strategies, which has dominated donor–government discourse in many countries in eastern and southern Africa, is the relative role of the state versus the private sector in any of the above policy options. Even if the private sector is the chosen *modus operandi*, how can the difficult transition from a highly interventionist environment to a liberalized one be managed without exacerbating the chronic food insecurity of the vulnerable? And how can the vulnerable be protected by government in such a way that the liberalization process is not thwarted?

The debate between government and donors in Malawi regarding food security strategies has, for decades, been dominated by the issue of the relative role of government and the private sector (Harrigan, 2003). Arguments over the future of the state-owned Agricultural Development and Marketing Corporation (ADMARC), the National Food Reserve Agency (NFRA) and who should manage the country's Strategic Grain Reserve (SGR) have diverted attention away from the more fundamental issue: how to tackle chronic food insecurity. A more sensible approach would be to develop a consensus around a viable food security strategy and then to work out the respective roles of government and the private sector in implementing the strategy, and how these roles might evolve over time.

17.3 Malawi's Food Security Strategy

17.3.1 Domestic maize production and complementary initiatives

In recent years, the government has emphasized the importance of Starter Pack, seeing it as part of a domestic maize self-sufficiency approach designed to reduce maize imports. However, it would be wrong to characterize government food security policy as focusing exclusively on maize production. Both government and external observers acknowledge that, in addition to raising maize output via the introduction of high-yielding, fertilized varieties, four other approaches require serious consideration (Devereux, 1997; Orr and Orr, 2001):

- First, food security needs to be improved by diversifying food crops towards more drought-resistant and less nutrient-demanding crops, such as cassava, sweet potatoes, groundnuts and beans. It is clear that such diversification is already under way in Malawi (see Chapter 12). However, Malawi does not produce enough non-maize staple foods to bridge the food gap. Moreover, estimating output of such foods is difficult, as was clearly illustrated in 2001/02, when over-estimates of cassava production by Sichinga *et al.* (2001) led to complacency and delayed response to the food crisis (Stevens *et al.*, 2002).
- Second, cash crop production should be promoted to improve food security both by earning foreign exchange for food imports and by increasing household purchasing power. In Malawi, this entails the promotion of traditional smallholder crops such as tobacco and cotton, as well as new high-value crops such as spices, fruit, vegetables and soybeans.
- Third, there is a need to promote growth in non-farm income as a route to livelihood diversification and improved food access (Orr *et al.*, 2001). This would involve the promotion of micro- and small-scale enterprises such as beer brewing, food processing and furniture making.[1]
- Finally, food production could be scaled up to more productive larger farmers, who would provide income-earning opportunities for poorer farmers in the form of wage labour. However, this strategy has not been considered seriously so far, as it is at odds with the Starter Pack approach (it relies on high maize producer prices, whereas Starter Pack's success depends on keeping maize prices low).

17.3.2 Imports and Strategic Grain Reserves

Malawian politicians and government officials have long been wary of a strategy that puts too much emphasis on food imports. A decade and a half of civil war in Mozambique, cutting off Malawi's most direct access to the sea, has engendered an almost siege mentality – which persists despite the ending of hostilities in Mozambique. Throughout the 1980s and early 1990s, large SGRs and fertilizer buffer stocks, partially financed by the European Commission (EC), were a reflection of this mentality. Even now, as Malawi is a landlocked country with poor transport infrastructure in neighbouring countries, the viability of extensive reliance on bulky food imports remains in doubt.

Attempts to deal with the 2001–2003 food shortages graphically illustrated the logistical problems of estimating and correctly timing food imports and dealing with transport bottlenecks, as well as the complex interactions between government imports and private-sector importers. Malawi is also acutely exposed whenever a regional food shortage occurs, since its larger neighbours – Mozambique, Zambia and Zimbabwe – have first call on food imports by virtue of their greater size and purchasing power and better connections with South Africa and overseas markets. Logistical problems aside, simple calculation illustrates that it is much more expensive for Malawi to import maize than to feed herself in terms of both direct costs and indirect macro-economic impact (see Chapter 15).

The United States Agency for International Development (USAID) believes that macro-economic problems can be avoided if the private sector is relied upon to import maize (Rubey, 2003). However, given the nascent nature of Malawi's private sector, the existence of information gaps and the government's desire to intervene whenever household food security seems threatened, it will be many years before the private sector can be relied upon to arbitrate markets and import food in a timely manner.

An SGR as part of a national food security strategy also presents problems. Management of such a reserve is difficult and costly, and the grain scandal of 2001 showed the scope for profiteering when releasing and selling the reserve (Stevens *et al.*, 2002). Decisions regarding quantities to be released and selling prices also have the potential to disrupt the market and undermine the activities of private traders.

Malawi's principal donors now seem to agree that a grain reserve of around 30,000–60,000 t is desirable (McMullin *et al.*, 2003), but this is based on the assumption that the main function of the SGR should be to support safety net interventions rather than to act as a 'buffer' in poor harvest years, when importing food at short notice might be problematic. Government officials who feel that the SGR should play an effective buffer (and seasonal price stabilization) role have argued for stocks in excess of 60,000 t.

17.4 Donor Views on Food Security Strategies for Malawi

The principal donors concerned with food security in Malawi are the EC, the UK Department for International Development (DFID), USAID, the World Bank and the International Monetary Fund (IMF).[2] In the mid to late 1990s, there was a policy vacuum in terms of food security strategies for Malawi, but in the past 4 to 5 years, donor views on viable strategies have evolved rapidly.

Historically, although government and donors acknowledged the need for livelihood diversification, the prime emphasis for both national and household/individual food security was on boosting domestic maize production. ADMARC subsidized input prices heavily and made a loss on its maize trading activities by offering high, guaranteed pan-territorial maize prices to producers whilst selling at loss-making, subsidized consumer prices.[3] Such policies were complemented by research and extension work to promote the substitution of local maize with higher-yielding hybrid varieties.

Apart from a brief period in the mid-1980s, when the IMF and the World Bank, as part of their Stabilization and Structural Adjustment Programmes in Malawi, requested the government to remove input subsidies and suppress the relative producer price of maize (Harrigan, 2003), most donors made little objection to the government's emphasis on maize production. In the mid to late 1990s, the World Bank and the EC reluctantly supported large-scale free maize seed and fertilizer distribution in the form of the Drought Recovery Inputs Programme (DRIP) and its successor, Starter Pack.

By the early 2000s, however, donors and government were starting to formulate a much broader food security strategy, de-emphasizing hybrid maize production. Despite this commonality, the government continued to see Starter Pack as an important part of its food security strategy, while most donors were increasingly sceptical.

17.4.1 The metamorphosis of the EC's position

Both the Malawi Government and the donor community acknowledge the EC's leading role within the donor community on food security issues. During the 1980s, the EC played a key role in funding the 180,000-t SGR as well as the importation, warehousing and running of the fertilizer buffer stock. More recently, its support for food security in Malawi has taken the form of a Multi Annual Food Security Programme (MAFSP).[4] According to McMullin *et al.* (2003), 'MAFSP has laid the foundations . . . for improved food security through a combination of leadership in the donor community, insistence on improved GoM policy and positive central institution building and capacity building.'

In February 2002, the government set up a National Food Crisis Joint Task Force with representatives of government, donors, UN agencies and civil society. This has been supported by an EC-funded technical secretariat of EU and Malawian experts based in the Ministry of Agriculture (MoA).

The EC is funding work by one of the Task Force's six sub-committees on drafting a National Food and Nutrition Security Policy (Government of Malawi, 2004a).

Given the dominance of the EC as a donor in this field, it is useful to trace the evolution of EC thinking regarding viable food security strategies in Malawi and the implications for Starter Pack. At the time when Starter Pack was launched in 1998, the EC, like the government, tended to focus on maize production. It supported Starter Pack (McMullin *et al.*, 2003), as well as the Agricultural Productivity Investment Programme (APIP), which provided smallholders with credit in kind (inputs) for producing maize, leguminous food crops and cash crops. However, work on the second MAFSP resulted in a belief that mirroring government policy had 'fault lines', and EC policy underwent 'fundamental changes' (McMullin *et al.*, 2003). The new approach, embodied in the second MAFSP, focused on livelihood diversification and food accessibility rather than food production and supply. The EC strongly believed that this gave it a wider poverty-reduction focus, orientated not just towards smallholders but also towards the vulnerable landless or near-landless poor.

Although it is difficult to pinpoint the impetus for this shift in policy, several factors seem to have contributed. First, the 2001/02 food crisis, which forced many donors to reappraise their approach to food security in Malawi and to acknowledge that the causes of chronic food insecurity are complex and diverse. Some concluded that the previous policies, including Starter Pack, had failed. Second, the influence of work on the government's Poverty Reduction Strategy Paper (PRSP) (Government of Malawi, 2001) in preparation for the IMF's Poverty Reduction Growth Facility and Highly Indebted Poor Country (HIPC) debt relief. This work elevated concerns about poverty in Malawi, identified its complex causes and advocated a social safety net approach as an important component of poverty reduction. Clearly, the new EC approach to food security meshed closely with the government's PRSP. Finally, in the words of one EC representative in Malawi, institutional changes within the EC meant that 'policy became less Brussels driven and more decentralized' (personal communication, Lilongwe, 2004).

17.4.2 The stance of other key donors

Given the EC's lead role on food security issues, it is not surprising that other key donors followed the EC's lead in the early 2000s by starting to articulate more holistic policies, influenced also by the evolution of the Malawian policy environment in the context of the PRSP. However, there were some important differences, resulting in a spectrum of views. At one extreme, DFID continued to believe that support for domestic maize production was important. At the other extreme, USAID had a vision of a completely monetized market economy where farmers diversify their livelihoods and use their cash income to access maize, much of which would be imported.

17.4.2.1 DFID

DFID has been characterized by other donors as failing to allow its food security policy to evolve and sticking to an emphasis on maize production via continued support of Starter Pack. Until recently, key personnel in DFID continued to see the promotion of improved maize via input provision as one important component of a viable food security policy.

However, personnel changes at DFID's Malawi office and work on the draft Growth and Livelihood Strategy in 2003/04 suggest that a new policy stance is beginning to emerge. As with the EC, this involves greater emphasis on livelihood diversification with safety nets to be provided via public works programmes (PWPs) – such as the fertilizer-and-seed-for-work scheme known as Sustaining Productive Livelihoods through Inputs for Assets (SPLIFA). This has implications for DFID's future commitment to Starter Pack, which some in the DFID-Malawi office now see as competing with the PWP approach to individual food security.

DFID's recently retired Livelihoods Adviser based in Lilongwe[5] focused on agriculture and the food sector. His lengthy development experience, including more than 8 years in Malawi, gave him a position of unique influence, much broader than that of most DFID advisers. Although the DFID office in Malawi has confirmed its commitment to continued support to national food security, the departure of such a key interlocutor is bound to affect how its contribution to national policy and programmes is handled. DFID also appears to be moving toward a more regional approach to food security in southern Africa, led by an adviser based in Pretoria. The future balance between the Lilongwe office and the regional programme and associated staffing is as yet unclear.

17.4.2.2 USAID

USAID is sometimes seen as lacking a clear food security strategy in Malawi. According to one observer (personal communication, Lilongwe, 2004):

> USAID has never bought into having a food security policy. Their position is that there are enough policies around already and that if they were implemented we wouldn't need an integrated food security policy.

To the extent that USAID has a long-term approach, this is clearly based on faith in the private sector. Although it would be unfair to characterize USAID policies in Malawi as driven exclusively by a pure free market stance (USAID supports many safety net projects, which inevitably distort markets), it does seem that the agency is more concerned than other bilateral donors with the potentially market-distorting effects of food security interventions. The faith in the private sector and the market mechanism has also led some in USAID to advocate a more radical food security position, which explicitly encourages a reliance on private imports from other countries in the region (such as Mozambique and South Africa) as well as a scaling down of the SGR. For instance, the leader of USAID's Sustainable Economic Growth Team in Malawi argues that decision makers should:

> Recognize that the single-minded pursuit of a policy of maize self-sufficiency is counter-productive and effectively condemns many Malawians to perpetual poverty. Malawi is likely going to be a net importer of maize in the future, but this need not be viewed as a failure . . . Past analyses have suggested that the costs of maintaining a grain reserve that would be large enough to make up for a significant shortfall . . . are simply too high and the potential benefits to the country too low (Rubey, 2003).

According to some government officials, USAID's pro-import stance has led to a degree of tension with government and the EC, with the latter arguing that if non-maize food crops are taken into account there is no reason why Malawi cannot be largely food self-sufficient.

17.4.2.3 The World Bank

Like the EC, the World Bank during the 1990s advocated free input distribution as a way of boosting maize production. In the 1992/93 season, the voucher-based Supplementary Inputs Programme, whereby ADMARC supplied seed and fertilizer, was entirely funded by the Bank. The Bank also supported the DRIP in the mid-1990s and Starter Pack in 1998/99 and 1999/2000 (see Table 17.1).

Nevertheless, the Bank acknowledged that, in the light of small land holdings, livelihood diversification was important (World Bank, 1995). In its eyes, a key element of such diversification was the promotion of smallholder burley tobacco production (Harrigan, 2003), which it was hoped would inject purchasing power into rural communities and kick-start the process of wider diversification. Some observers, however, feel that the Bank has failed to develop a clear position on food security strategy in Malawi.

Even if it lacks a clear policy framework, the World Bank has moved increasingly towards an emphasis on social safety nets for overcoming chronic food insecurity. It was influential in recasting the original Starter Pack programme as the Targeted Inputs Programme (TIP) as part of the National Safety Net Strategy (see Chapter 2).

17.4.2.4 IMF

The IMF is a key player in Malawi and many other developing countries, in that an on-track IMF programme is a signal for other donors to release budgetary support. The IMF agreement of December 1995 explicitly advocated agricultural market liberalization and a removal of input subsidies. In the PRSP, negotiated with the IMF, the Malawi Government committed itself to maintaining competitive conditions in grain and fertilizer markets.

In 2001, the IMF explicitly intervened in food security issues by asking the government to reduce the SGR from 167,000 to 60,000 t through domestic sales and exports. This was to enable the repayment of a US$300 million foreign commercial loan by the NFRA when it was established as a quasi-independent agency (Owusu and Ng'ambi, 2002). Despite an impending food crisis, the government proceeded to sell almost all of the 167,000-t reserve to private traders, who stockpiled in a manner which both exacerbated the subsequent hunger and enabled them to benefit from it. The sales were beset by

Table 17.1. Financing of Starter Pack and the TIP (US$ million).

Year	Malawi Government	DFID	World Bank	EC	NORAD	IFAD	Republic of China	Libya	Total	Malawi Government (% of total)
1998/99	14.5	8.2	1.7	0.7			0.5		25.6	56.6
1999/2000	12.5	4.3	7.2			1.2			25.2	49.6
2000/01	2.3	2.8		1.5				0.9	7.6	30.3
2001/02	2.4	3.6		1.2					7.2	33.3
2002/03	1.3	9.9			2.2				13.4	9.7
2003/04	1.2	10.9							12.1	9.9
Total	34.2	39.8	8.9	3.5	2.2	1.2	0.5	0.9	91.1	37.5

Source: Starter Pack/TIP Logistics Unit records.

allegations of mismanagement and corruption (Government of Malawi, 2004b).

The IMF received much criticism for its advice to the government on reducing the SGR. In August 2002, it authorized the government to subsidize consumer prices for the 2002/03 season in view of the continuing food shortages – despite substantial cost implications for the Treasury (see Chapter 15). At the same time, it distanced itself from food security issues, deferring instead to the World Bank:

> The World Bank has served as the lead advisor on agricultural and food security policy reform. Given the impact of agriculture on the budget, however, certain elements of the reforms were supported under recent IMF arrangements as well (IMF, 2002).

These elements included attempts to end government and ADMARC interventions in maize markets, to develop new operational guidelines for the NFRA and to keep maize stocks in the SGR at relatively low levels (see Section 17.3.2).

However, the IMF's position is principally one of disengagement from food security policy. It believes that a hands-off approach will also help to smooth over the damage caused to its reputation by the 2001 SGR affair. Thus, despite the tight fiscal position, it raised no objection to the Malawi Government including expenditure of US$22 million for subsidizing fertilizer distribution in its 2004/05 budget.

17.5 Donor Positions on Starter Pack and TIP

The general view of Malawi Government officials involved in the design of the first Starter Pack was that the donor community was, at worst, hostile and, at best, lukewarm. The exception was DFID, which announced that it was willing to support the programme until someone devised a better intervention to tackle chronic food insecurity. Table 17.1 shows that the only consistent financial supporter of the programme among donors has been DFID. DFID's support has also been critical in terms of getting commitment to the programme from government and other donors.

Given the subsequent World Bank opposition to Starter Pack, its initial support might seem surprising. Some analysts have interpreted the Bank's reluctant support in a cynical light, arguing that it was motivated by pragmatic self-interest. The Bank was US$6 million under-spent on its Agricultural Services Programme (ASP) in the late 1990s, and Starter Pack provided a convenient opportunity to disburse these funds. Second, the Bank was pushing hard for agricultural input market liberalization. In 1998, the government had 30,000 t of fertilizer stocks, which could be unleashed on the market at any time, threatening to undermine commercial pricing policy and market development. Hence, according to the Head of the Starter Pack/TIP Logistics Unit:

> The donors had a hidden agenda. Starter Pack was a controlled way of getting rid of the government's fertilizer stock. That's why they were happy for it to be called 'Starter' Pack – they only envisaged it as a short-term programme, but the problem was that government got hooked (C. Clark, Lilongwe, 2004, personal communication).

EC support for Starter Pack in 1998/99 and the TIP in 2000/01 and 2001/02, which largely took the form of seed provision and cash for logistics, seems to have been motivated by the fact that during this period the EC's own food security policy was only slowly evolving away from one which emphasized maize production. After the second MAFSP came on stream, with its emphasis on livelihood diversification (see Section 17.4.1), the EC withdrew support from the TIP.

Four other donors have made single-year financial contributions to Starter Pack/TIP (see Table 17.1). This does not, however, seem to have had much to do with their support for the underlying principles of the programme. The Norwegian aid agency, NORAD, for example, contributed US$2.2 million in 2002. However, this was because it had spare funds as a result of the inability to disburse budget support at a time when Malawi was off-track with its IMF programme. As the country was in the midst of a food crisis, the TIP seemed to be a good way of using the funds.

Table 17.1 shows that in 1998/99 and 1999/2000, the government contributed half of the total cost of the Starter Pack programme (partly via the use of the government fertilizer stock). However, the government's contribution fell to 10% by 2002/03. This was due to IMF pressure to restrict government expenditure on the TIP to under US$2.5 million as part of the fiscal austerity programme.

The higher share contributed by the donors gave them more leverage over the design of the programme. Within a few years, donors were articulating a wide range of objections to the original Starter Pack programme, which coincided with the evolution of their food security policies as well as an elevation of the social safety net programme in Malawi. Starter Pack became the TIP largely due to donor pressure (see Chapter 2)[6]. In the words of one donor representative:

> From 2000 onwards, there was a meeting of minds between the EC, USAID and the Bank. We didn't see Starter Pack as a long-term strategy, even if government did, and we started to develop an exit strategy towards more targeted interventions (personal communication, Lilongwe, 2004).

As time passed, donors increasingly argued that the TIP was inefficient as a social safety net and that self-targeting PWPs such as food for work, inputs for work and cash for work were more appropriate schemes.[7] According to McMullin *et al*. (2003):

> Some see TIP which provides free limited agricultural inputs as a catalyst to help farmer households to generate surpluses and higher demand for a wide range of goods. While others, including the EC Delegation team, opt for the potential of labour intensive public works projects to provide the impetus towards surplus generation.

At the same time, donors felt that PWPs should be complemented by the development of input markets to improve input uptake by farm households. The main engine for the latter was the USAID-supported Agricultural Input Markets Development Project (AIMS), which attempts to improve smallholder access to improved seeds, fertilizer and crop protection products through the promotion of private-sector suppliers.

Donor criticisms of the Starter Pack and TIP also covered perceived operational weaknesses. For instance, they pointed out that the pack content is inappropriate for some geographic areas, that the inputs are wrongly used due to lack of extension advice and that distribution has sometimes been delayed. Some also argue that the programme has been politicized and is open to corruption. The World Bank in particular has been vocal in the view that TIP has been abused and poorly distributed. While these are indeed weaknesses of the programme, they are issues of programme design and implementation and do not constitute a fundamental criticism of the Starter Pack approach.[8]

Another operational concern about the TIP is that it did not succeed in targeting the poorest households (see Chapter 11). On the other hand, there is a degree of donor inconsistency here in that many donors also argue that if the TIP is designed to boost productivity it should be targeted at the wealthier households who can use the packs more productively. These inconsistencies seem to reflect a lack of clarity about the programme's aims and objectives (i.e. whether it is a production-enhancing intervention or a safety net).

Creation of a 'dependency syndrome' among smallholder farmers is another oft-cited criticism of free inputs distribution, despite the evidence to the contrary (see Chapter 9). The TIP is often compared with PWPs, which are seen as less dependency-creating because inputs are exchanged for work rather than handed out free. The dependency syndrome is often expressed by donors in terms of 'the need for an exit strategy from the TIP' at household and national levels.

Starter Pack and TIP are also criticized for reinforcing the country's dependence on maize and inhibiting diversification into non-maize food crops and cash crops. In the words of one donor representative it 'reinforces the maize trap': the dominance of maize in the diet means that government and donors wrongly focus attention on maize, which in turn reinforces the dominance of maize and condemns Malawians to a cycle of poverty (Rubey, 2003). In this respect, donors often point to cash-for-work PWPs as preferable, as they provide beneficiaries with flexibility in terms of potential input purchases. PWPs which improve rural infrastructure, for example roads and irrigation, can also help promote crop and livelihood diversification.

One of the most widely articulated donor criticisms of Starter Pack and TIP is that they undermine the development of private-sector input markets. It is argued that the programme is centrally managed, and although transport, warehousing and distribution are contracted out, it is the larger urban-based operators who benefit, squeezing out rural retail outlets. Related to this is the argument that the programme undermines other projects attempting to develop private-sector input markets such as AIMS, which works with the National Smallholder Farmers' Association of Malawi (NASFAM) and SPLIFA.

17.6 A Critique of the Donor Position

Without prejudging the question of whether Starter Pack is a good programme for tackling chronic food insecurity in Malawi, we can identify some weaknesses in donors' positions. These weaknesses have inhibited a constructive and objective debate as to the future of the programme.

17.6.1 The wrong questions

The Starter Pack debate has often been derailed by questions of 'who does what' rather than 'what should be done?' Rather than evaluating the programme as a set of actions designed to address food insecurity, the debate has tended to focus on Starter Pack as 'a government project that undermines the private sector'. This preoccupation with the relative roles of government and the private sector has diverted attention away from the key issue of whether Starter Pack has a role to play in the country's food security policy.

Similarly, another common donor question: 'Does Starter Pack promote maize production to the exclusion of other crops?' is not the correct question. A more useful question is: 'How best can the maize diet be enriched?' In this respect, maize and other food crops can be seen as complementary without continued smallholder maize production necessarily inhibiting evolution towards food-crop diversification. Indeed, Starter Pack, which combines maize seed with legumes, takes such an approach (see Chapter 12).

17.6.2 Misinterpretation

Related to this is the problem of donor misinterpretation of the original Starter Pack concept. Despite the perceptions of most donors, the programme was never seen by those who designed it, implemented it or evaluated it as an alternative to a more holistic food security and livelihood diversification policy. Rather, it was regarded as one part of such an approach (see Introduction and Chapter 1).

Likewise, it is wrong to characterize the government as having deliberately introduced a programme which it knew would undermine private input markets. The intention was quite the reverse: to provide inputs to those who would not normally have bought them so as to stimulate purchases of inputs from commercial suppliers.

The alteration of the programme from Starter Pack to TIP led to confusion over the programme's objectives. However, it is wrong to criticize the government for running an ineffective social safety net programme. This was never the government's objective, and the move in this direction was largely donor-driven.

17.6.3 Lack of evidence-based critique

Despite receiving annual evaluation reports (see Chapters 5 and 6), donors' positions on Starter Pack and TIP often seem to have been based on an ideological reaction, rather than on concrete evidence. This is particularly true of their claim that the programme undermines private input markets. The Starter Pack/TIP Monitoring and Evaluation (M&E) programme found little evidence of this (see Chapter 10), and many of Malawi's fertilizer distributors supported the free inputs distributions. As one person involved in the design of the programme observed (personal communication, Lilongwe, 2004): 'The private sector was involved and didn't feel they were being crowded out. The EC and USAID didn't talk to them to gather evidence on this.'

Lack of an evidence-based donor position is also shown by the failure to realize that the Starter Pack/TIP programme has evolved since 1998. Although in the early years the Logistics Unit relied heavily on government and related organizations for the purchasing, warehousing and distribution of the inputs, more recently, concerted efforts have been made to involve more private-sector players via a competitive tendering process (see Chapter 3). According to the Head of the Logistics Unit, although 'the programme has evolved in ways that answer some of the donors' concerns . . . donors don't seem to have acknowledged any of this' (C. Clark, Lilongwe, 2004, personal communication).

More generally, the donor view that the Starter Pack programme inhibits reform is at odds with some of the evidence. For example, in 1998/99 and 1999/2000 Starter Pack helped produce bumper maize harvests. Many in government believe that it was only because of this that they were able to press ahead with reforms that donors had been demanding for years: dismantling ADMARC, removing the maize price band and rescinding the ban on maize exports. Subsequent maize deficit years have resulted in the reversal of many reforms.

Finally, it is surprising that members of the donor community hardly mention the scaling down of Starter Pack in 2000/01 as one of the probable causes of the 2001/02 food crisis (see Introduction). This is strongly at odds with the position taken by Malawi's president, Bakili Muluzi, in May 2002, who claimed that: 'Problems started when donors said we should cut the number of recipients for the Starter Pack from 2.3 million to one million in the Targeted Inputs Programme' (AllAfrica.com, 2002).

17.6.4 Lack of properly costed alternatives

'Is Starter Pack affordable?' is another case of asking the wrong question. According to Levy *et al.* (2004), the correct questions should be 'How expensive is it compared to credible alternative food security interventions?' and 'What is the cost of not implementing it in terms of food imports and associated macro-economic instability?' Although donors have been quick to criticize the programme as fiscally unsustainable, they have failed to carry out rigorous analysis of its costs and benefits compared with other alternatives such as input subsidies and/or maize imports (see Chapter 15).[9] In the words of a former adviser to Malawi's Vice President:

> US$25 million on Starter Pack got the food problem under control. This was a small cost compared to imports, but donors didn't look at the costs of alternatives. In 1999/2000 we were fine food-wise but no one bothered to sit down and work out why we were fine. If they had, they would have found that Starter Pack played an important role (A. Conroy, Lilongwe, 2004, personal communication).

On a more general level, the debate as to whether Malawi should concentrate on cash-crop production and move towards more maize imports has not been informed by the kind of analysis that can quantify the relative merits of this strategy compared to domestic maize production.[10]

17.6.5 Donor in-fighting

Finally, perhaps the most alarming aspect of the donor community's position on Starter Pack and food security is the degree of in-fighting. For example, in recent years the EC has been hostile to DFID's support for the TIP, while on the other hand, many donors and occasionally government officials see the EC delegation as 'driving its own agenda' via the large amount of resources it has for food security interventions. Constructive donor discussion on the future of Starter Pack has been prevented by the view in the rest of the donor community that it is 'a DFID project'. As one observer remarked: 'The EC/DFID relationship has been quite confrontational over TIP. Also, others saw TIP as DFID's project and so didn't engage with it' (personal communication, Lilongwe, 2004).

17.7 Conclusion

Starter Pack – a government initiative that had a substantial impact on food production – promised to be an excellent example of 'ownership', 'partnership' and 'aid effectiveness', all of which are currently buzz words in the donor community. However, from the outset, with the exception of DFID, it was met with little donor support or enthusiasm.

The main effect of donor engagement with Starter Pack was to change it from a production intervention to a targeted social safety net in the form of TIP. As donor food security policies evolved, donors saw less room for the programme. Yet many government officials, although keen to promote the livelihoods and accessibility dimensions of food security, remain committed to Starter Pack. How do we reconcile these two positions? Perhaps the answer lies in differences in time horizons and attitudes to risk. Many donors articulate an essentially long-term vision of food security in Malawi, involving diversified food production, increased smallholder cash-crop production and off-farm income, and well-developed private markets, enabling individuals to utilize their increased purchasing power to access food. In such an environment, they see no need for handouts of free inputs except perhaps as a very limited form of safety net for the most vulnerable. Most people in government share this vision of the future, but are acutely aware that it is a long-term vision, which will be preceded by a protracted and risky transition phase. Government officials are concerned about the food security risks during this transition, and see Starter Pack as having a role to play as a domestic maize-production-enhancing intervention.

Policies for transition pose considerable policy dilemmas in terms of avoiding chronic food insecurity. What role should the government play? Should there be a focus on maize production? Is it sensible to rely on maize imports? Should there be free inputs, input subsidies or simply efforts to develop private input delivery? These and many more questions need to be addressed and policies need to be devised that can smooth the transition and ensure that it does not jeopardize food security. Unfortunately, weaknesses in the donor response to the Starter Pack programme, including lack of an evidence-based position, in-fighting, misinterpretation, posing of the wrong questions and failure to produce properly costed alternatives have prevented a sensible dialogue on what role, if any, Starter Pack might play in this difficult transition phase. It is to be hoped that this book will go some way towards prompting a more meaningful debate on the issue.

Notes

[1] In terms of a strategy of diversifying household livelihoods, either through cash-crop production or off-farm income, problems include: diverting land and labour away from food-crop production (necessitating either imports or promotion of large-scale food production units); finding a way for asset- and land-poor households to engage in livelihood diversification; and ensuring that markets for both cash crops and off-farm products exist and that individuals can use their purchasing power to purchase food with confidence.

[2] Although the IMF does not have explicit views on the country's food security strategy, it is nevertheless concerned with the macro-economic implications of policies in this area, whilst its policy advice regarding the exchange rate, interest rate and public expenditure has an impact on food security.

[3] Throughout the 1970s, ADMARC cross subsidized its maize and fertilizer trading account losses with some of the profits made on the purchase and sale of smallholder cash crops, especially tobacco, cotton and groundnuts, for which producers were paid prices well below farm gate export parity level (see Harrigan, 1988, for a more detailed historical analysis).

[4] The first MASFP ran from 1998–2002 with a budget of 32.4 million euros, and the second from 2001–2004 with a budget of 53 million euros; a third for 2004–2007 is currently being designed.

[5] Harry Potter, the author of Chapter 2.

[6] The only international agency that continued to support the TIP as a maize-production-enhancing initiative was the UN's World Food Programme.
[7] In 2001, the EC started supporting the Employment Generation Scheme (EGS), a food security related PWP. The SPLIFA, an inputs-for-work PWP (see Section 17.4.2), has been running on a pilot basis for 2 years with funding from DFID.
[8] Other input distribution programmes such as APIP have also suffered logistical problems, and late distribution of Starter Pack/TIP is partly a reflection of donor unwillingness to make long-term commitments to the programme, so it ends up being put together in a hurry each year.
[9] Compared to many donor-funded projects, Starter Pack/TIP seems to be a good example of aid effectiveness. The World Bank's assessment of its projects in Malawi during the 1990s found that only 57% had satisfactory outcomes and 15% were likely to be sustainable (World Bank, 2000).
[10] For example, the Policy Analysis Matrix is a tool that can be used to calculate the opportunity cost of using domestic resources (measured at world prices) to produce different crops. This can help policy makers decide whether it is economically efficient to generate foreign exchange via crop exports and import food (see Harrigan et al., 1992).

References

AllAfrica.com (2002) US government provides food aid. Available at: http://allafrica.com

Devereux, S. (1997) Household food security in Malawi. *IDS Discussion Paper* No. 362. Institute of Development Studies, Brighton, UK.

Government of Malawi (2001) Malawi poverty reduction strategy paper, National Economic Council, Lilongwe, Malawi.

Government of Malawi (2004a) Draft food and nutrition security policy. Ministry of Agriculture, Lilongwe, Malawi.

Government of Malawi (2004b) Report of the presidential commission of inquiry on strategic grain reserves under the chairmanship of Khuze Kapeta, Lilongwe, Malawi.

Harrigan, J. (1988) Malawi: the impact of pricing on smallholder agriculture 1971–1988. *Development Policy Review* 6, 415–433.

Harrigan, J. (2003) U-turns and full circles: two decades of agricultural reform in Malawi 1981–2000. *World Development* 31(5), 847–863.

Harrigan, J., Loader, R. and Thirtle, C. (1992) Agricultural price policy: government and the market. Training Materials for Agricultural Planning, No. 31, FAO, Rome.

IMF (2002) Malawi – the food crisis, the strategic grain reserve, and the IMF. *IMF Factsheet*, Washington, DC.

Levy, S. with Barahona, C. and Chinsinga, B. (2004) Food security, social protection, growth and poverty reduction synergies: the Starter Pack programme in Malawi. *Natural Resource Perspectives* No. 95. Overseas Development Institute, London.

McMullin, P., Antoniou, N. and O'Leary, M. (2003) Mid-term evaluation of the European Commission Food Security Programme 2001–2004 in Malawi. Transtec and AFCon, Lilongwe, Malawi.

Orr, A. and Orr, S. (2001) Changing livelihoods in Malawi's rural south: a synthesis and interpretation. Kadale Consultants, Blantyre, Malawi.

Orr, A., Mwale, B. and Saiti, S. (2001) Market liberalisation, household food security and the rural poor in Malawi. *The European Journal of Development Research* 13(1), 47–69.

Owusu, K. and Ng'ambi, F. (2002) Structural damage: the causes and consequences of Malawi's food crisis. World Development Movement, London.

Øygard, R., Garcia, R., Guttormsen, A., Kachule, R., Mwanaumo, A., Mwanawina, I., Sjaastad, E. and Wik, M. (2003) The maze of maize: improving input and output market access for poor smallholders in southern Africa region, the experience of Zambia and Malawi. Agricultural University of Norway, Department of

Economics and Resource Management, Report No. 26.

Rubey, L. (2003) Malawi's food crisis: causes and consequences. USAID, Lilongwe, Malawi.

Sichinga, K., Salifu, D. and Malithano, D. (2001) EC food security and food aid programme crop assessment report 2000/2001, Vol. 2, final report. Agricultural Policy Research Unit, Bunda College, University of Malawi, Malawi.

Stevens, C., Devereux, S. and Kennan, J. (2002) The Malawi famine of 2002: more questions than answers. Institute of Development Studies, Brighton, UK.

World Bank (1995) Malawi agriculture sector memorandum: strategy options in the 1990s. Report No. 12805 MAI, World Bank, Washington, DC.

World Bank (2000) Malawi country assistance evaluation, World Bank, Washington, DC.

18 Feeding Malawi from Neighbouring Countries

Martin J. Whiteside

18.1 Introduction

Malawi is a landlocked country of relatively high population density, intense land pressure and household food production shortfalls. It is surrounded by parts of Mozambique, Tanzania and Zambia with much lower population densities, more available land, and existing or potential food production surpluses.

Maize flows from the surrounding countries, particularly northern Mozambique, to markets in Malawi. However, much of this movement is informal and does not enter government statistics. Depending on the year, regional maize trade is thought to make up between 5% and 25% of Malawi consumption. It has an important impact on Malawi's food security. It is also a key economic driver, particularly for northern Mozambique.

Cross-border trade raises a number of interesting questions:

- To what extent can such trade plug food gaps at national and household levels in Malawi?
- Are there ways to predict informal trade flows so as to be able to predict and manage food deficits (or surpluses)?
- What is the impact of interventions like Starter Pack on trade and on the livelihoods of farmers in surrounding countries? Do production subsidies in one country harm farmers in another?
- What strategies maximize benefits for both neighbouring farmers and Malawian consumers?

18.2 Tradable Surplus in Neighbouring Areas

By far the most important traded food commodity for Malawi is maize, although smaller quantities of beans, cassava, sweet potatoes and European potatoes, sugar and meat are traded. The important production zones for maize in neighbouring countries are (Whiteside *et al.*, 2003):

Northern Mozambique. The provinces of Tete, Zambezia and Niassa surround the southern region of Malawi and are generally maize surplus producers. North–south transport links in Mozambique are poor and expensive: it is generally uneconomic to transport a northern surplus to southern Mozambique, even when there is a deficit in the south (Mozambique Ministry of Industry and Trade, 2001). Therefore, there is a tendency to export the surplus from northern Mozambique to Malawi. Maize production in northern Mozambique is around 700,000 t. Since cassava is a major staple food in northern Mozambique, maize is a cash crop for many households. Around half the maize crop (350,000 t) is potentially available for sale in most years. Mozambique has tended to resist calls for maize export bans, even in years of shortage (SIMA, 2002).

Malawi is the key market for northern Mozambican maize. However, farmers experience a boom and bust market: in some years there is market frenzy, with traders trying to buy all available maize, and in other years there is little demand. Mozambican farmers increasingly have a choice of cash crop – with cotton, tobacco, oilseeds and some pulses being potential competitors to maize. Therefore, the quantity of maize available to Malawi in future is likely to depend on perceived market demand, conditioned by farmers' experience of the market in preceding years.

South-western Tanzania. The regions of Iringa, Mbeya, Rukwa and Ruvuma (the southern Maize basket) are major maize producers. Typical maize production in these four regions of Tanzania is around 900,000 t, with a combined cereal surplus of around 300,000 t (most of it maize). However, unlike Mozambique, there are competing destinations for this maize: deficit areas of Tanzania, the Democratic Republic of Congo (DRC) and occasionally Zambia and Kenya. Moreover, Tanzania has only a short border with the far north of Malawi, increasing transport costs and making informal trade easier to control. In years of domestic shortage, Tanzania tends to impose an export ban, which is difficult to circumvent.

Zambia. Zambia's eastern and northern provinces abut the northern and central regions of Malawi. Northern province soils tend to have low fertility and production surpluses are small. There is more potential for exports from Zambia's eastern province and it is also closer to the major population centres of Malawi. Production levels in the eastern province in recent years have been around 240,000 t, but the exportable surplus seems to be fairly small.

18.3 Supply, Demand and Prices

Overall, therefore, northern Mozambique is the most promising source of supply. Tanzania is likely to be important in years of serious deficit

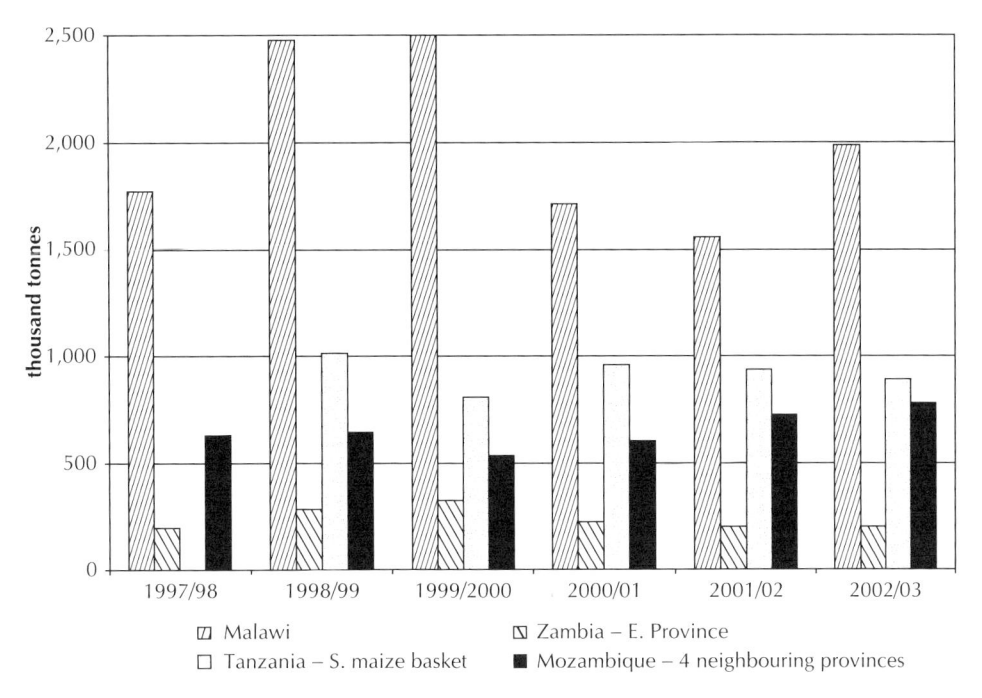

Fig. 18.1. Maize production in Malawi and neighbouring areas. (Calculations by Whiteside *et al.* (2003) based on data from the Ministries of Agriculture of Malawi, Mozambique, Tanzania and Zambia, FEWS and interviews with key informants.)

and high prices in Malawi, while Zambia's contribution is likely to be more modest. A comparison of these potential 'maize basket' areas with Malawi shows that Malawi currently out-produces them all (see Fig. 18.1).

However, more important than overall production levels are levels of deficit and surplus. In most years there is a maize deficit in Malawi. In years of severe shortage, Malawi's domestic maize prices rocket as wealthier rural households try to buy maize, exerting enormous demand pressure in the domestic market and on Mozambique, Tanzania and Zambia. Informal maize imports from Mozambique may reach 250,000 t in severe deficit years, while they may be as low as 70,000 t in years of good domestic supply.

The main driver for the cross-border maize trade is the price differential. Malawi, with its generally large consumption deficit, sets prices in the provinces of Mozambique, Tanzania and Zambia that border Malawi (see Fig. 18.2). In 2001/02, strong demand in Malawi drove up prices in neighbouring areas, reflecting a poor harvest in Malawi, average-to-poor production in the 'supply' areas of Mozambique, Tanzania and Zambia, an export ban in Tanzania and competition from Zambia and DRC for Tanzanian maize. It is important to note that the price levels in the 2001/02 season were beyond

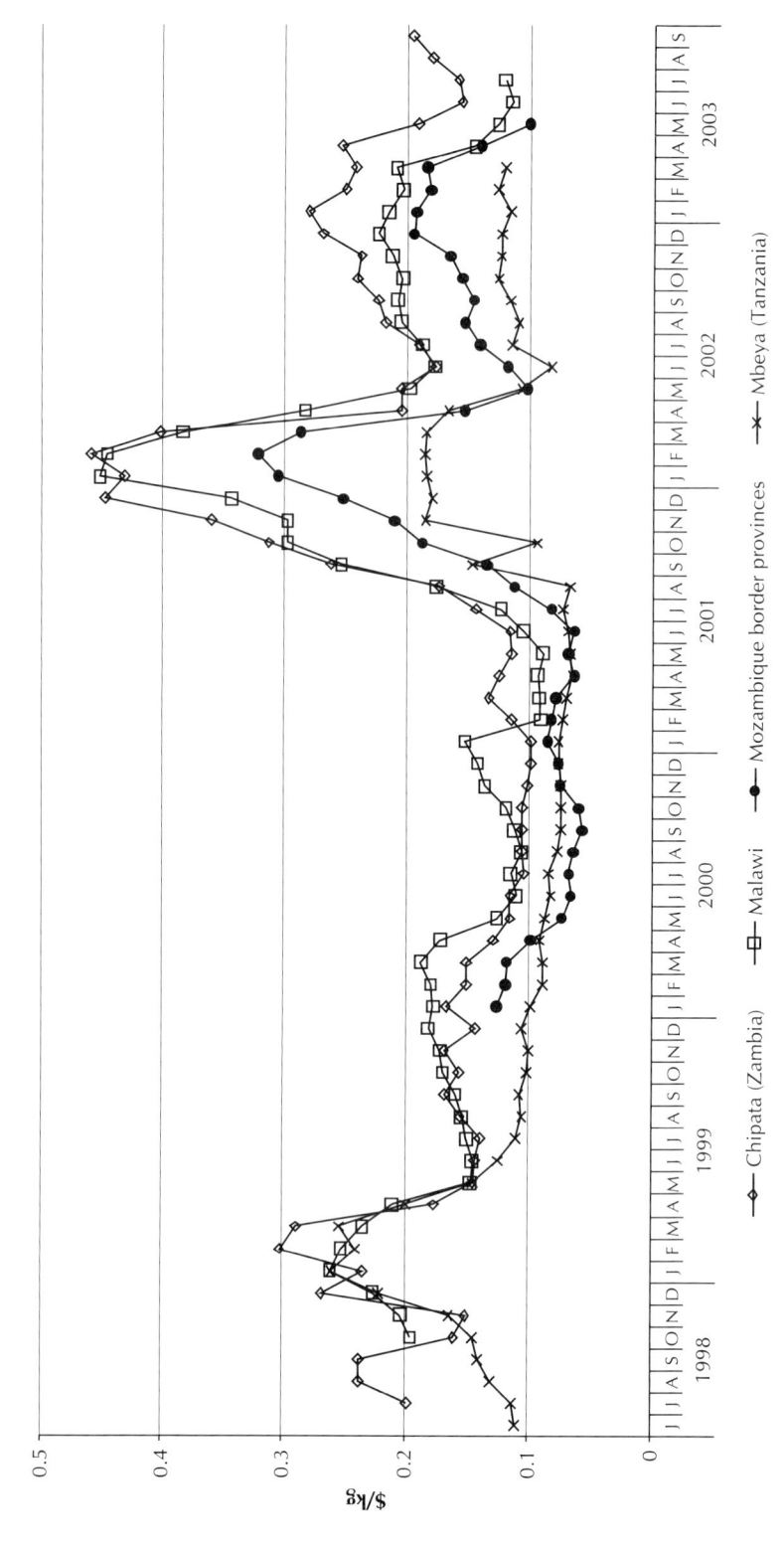

Fig. 18.2. Maize price comparison – Malawi and neighbouring areas. (Calculations by Whiteside et al. (2003) based on data from the Ministries of Agriculture of Malawi, Mozambique, Tanzania and Zambia, FEWS and interviews with key informants.)

the reach of most Malawian households, particularly rural consumers (see Chapter 8). Thus, while traders and farmers with a surplus – particularly in neighbouring countries – reaped the benefits of high maize prices in Malawi, the majority of Malawian farmers suffered the effects of food insecurity.

Malawi's 2002 maize harvest was even lower than that of 2001, but prices did not rise as much due to a combination of factors:

- better harvests in Mozambique and Tanzania;
- no export ban from Tanzania; and
- the arrival of large-scale maize imports and food aid organized by the Malawi Government, the World Food Programme (WFP) and aid agencies.

For exporters to Malawi, the rate of depreciation of the Malawi Kwacha (MK) is as critical as the domestic price of maize. Rapid depreciation of the MK (see Introduction and Chapter 15) creates a disincentive to farmers and traders in neighbouring countries wishing to export maize to Malawi. For years, much of the border trade in Mozambique has been conducted in MK, and Mozambican farmers have purchased consumer goods in MK through Malawian retail networks; but depreciation has undermined confidence in the MK among traders and farmers in neighbouring countries and there is reluctance to hold large amounts of cash in MK. For those who convert export sales revenue in MK into their own currencies (or into dollars), rapid depreciation of the MK means that the Malawi domestic price of maize needs to rise equally rapidly to prevent an erosion of the value of their exports. However, unless Malawi wages and incomes also rise rapidly, this inflationary process undermines demand for imported maize in the Malawi market.

Unfortunately for Malawi, neighbouring-country farmers are unlikely to produce a consistent surplus unless there is a consistent market (i.e. stable demand and sufficiently high prices). This is particularly true of northern Mozambican farmers, whose market options for surplus maize apart from Malawi are limited, and who suffer greatly if Malawi does not buy their surplus.

18.4 The Dynamics of Regional Trade

The maize trade between Malawi and its neighbours involves a variety of players. At the lowest level, it can involve a Mozambican border-area farmer carrying his or her maize to the nearest market, which happens to be in Malawi, and selling it there. Alternatively, Malawi border-area consumers may walk into Mozambique and buy from farmers or from local markets, or Malawians may do *ganyu* in Mozambique and get paid in maize.

At another level, informal-sector traders buy from farmers or markets at varying scales, but cross the border unrecorded before selling to Malawian markets or wholesalers. Often this involves unloading lorries on one side of the border, taking the maize across the border by bicycle, head-load or canoe and then reloading it on to a lorry on the Malawian side.

At the next level up, the maize may cross the border officially by lorry, but sometimes the documentation or phytosanitary certificates may be incomplete, false or inadequately recorded at the border. Finally, part of the trade enters with complete documentation and accurately enters national statistics.

The proportion of trade that is accurately recorded in official statistics varies from year to year and from border crossing to border crossing. Although there have been several attempts to estimate overall flows, the figures remain uncertain (Minde and Nakhumwa, 1997; Macamo, 1998; Whiteside, 1998, 2002; FEWSNET, 2002; Whiteside *et al.*, 2003).

The dynamism of commercial trade is extraordinary, particularly at the small trader level. Traders (often women) with low levels of formal education move maize around in a way that is highly responsive to changing profit margins – and with very low overheads. This sector is fairly new, as it is only in recent years that maize markets have left state control and are being deregulated (to different extents in the different countries). Mobile phone networks are increasing the potential for information exchange.

18.4.1 Barriers to trade

Competition among traders is fierce. The very diversity of the traders involved – different nationalities, with operations ranging in size from the person using his/her head or bicycle to large-scale operators with a fleet of trucks and network of warehouses – tends to mitigate against cartels forming, despite occasional attempts by larger traders in alliance with local officials. There are, however, barriers to trade that increase transaction costs, lowering prices to farmers and/or raising prices to the Malawian consumer. The main barriers are:

- *Weak infrastructure*, particularly poor roads in rural areas and near the borders.
- *Charges for vehicles crossing borders*. Various road taxes and insurances mean it is often cheaper to unload and load onto a different lorry when crossing a border – but this also increases costs.
- *Bureaucracy*. Although it is possible to trade legally, this often means getting the correct paperwork from distant administrative centres and waiting for approvals. However, risk taking and bribes associated with operating illegally also raises costs. Phytosanitary procedures can be cumbersome – yet it seems unlikely that they are sufficiently rigorous to prevent the spread of pests and diseases.
- *Corruption*. Even when cross-border trade is legal and tariff free, officials often expect payments.
- *Government-created uncertainty*. Maize is seen as a strategic commodity. Periodic and unpredictable government interventions in the market – buying or selling, setting prices, banning imports or exports, devaluation – create extra risk for traders, which raises costs. The intervention itself

may be less of a problem than the uncertainty: there is a need for greater transparency and predictability.

18.4.2 Measuring unrecorded trade

Failure to estimate informal imports was catastrophic in 2002/03, when the Malawi Government imported large quantities of commercial maize. The combination of food aid (reducing demand), and formal and informal imports (increasing supply) caused the price of maize to fall below forecast levels, causing serious fiscal damage (see Chapter 15). Traders and farmers in neighbouring countries saw their market collapse.

Various attempts are being made to estimate unrecorded trade, but estimates remain rough and there is little forecasting capacity (FEWSNET, 2002; SIMA, 2003; RATES, 2003a, b; Whiteside *et al.*, 2003). More work is needed on improving the estimates and developing a system for predicting informal imports using criteria such as production in supply areas, as well as making forecasts of price differentials and exchange rate changes.

18.5 Working with Cross-border Trade to Improve Food Security

A fundamental challenge for food security managers is to develop capacity to predict probable commercial food imports (recorded and unrecorded) to enable them to recommend appropriate government and donor action. This requires an understanding of the factors that drive the trade and an ability to predict how these factors are most likely to develop during the year. The key steps in the process are:

1. *Estimating Malawi's food balance for the coming year.* The domestic food balance is likely to be the primary driver in developing price differentials with neighbouring areas. However, the gross national balance needs to be complemented with more area- and household-specific vulnerability assessments. Moreover, the food balance estimation process is still quite crude. Production and consumption estimates are not robust, and there are particular problems with root and tuber crops.
2. *Look at price differentials, including the impact of exchange rate movements.* Is there likely to be an incentive to import? And from which source areas? Traders often have fairly accurate opinions on 'what prices will rise to this year'. However, exchange rate movements are more difficult to forecast.
3. *Make estimates of saleable surplus in neighbouring countries* in source areas with sufficient price differentials. This requires using crop production estimates, which are published at various stages by countries in the region. These can be triangulated with information from traders and from border posts.
4. *Check for changes in barriers to trade and competition from other countries.* Export bans are likely to reduce volumes considerably between Malawi and Tanzania but to have a lesser effect between Malawi and Mozambique and an

intermediate effect between Zambia and Malawi. Also, if demand develops in other markets, there may be competition for the surpluses produced by the identified source areas.

5. *Predict market-driven imports* – including recorded and unrecorded trade. This involves a number of iterative steps: Having calculated the preliminary food deficit, look at the probable surplus in those areas closest to the deficit areas, with low transaction costs and good price differentials. If these areas look unlikely to be able to meet the deficit, are there potential surpluses elsewhere? How much would prices need to rise for commercial supplies to start to flow?

6. *Answer the key questions*:

- Will markets fill the gap?
- How much are prices likely to rise?
- Is the predicted price rise acceptable in Malawi?
- Is there a need for government intervention? If so, what type?

7. *Monitor the situation throughout the season* – observing border flows and changing price differentials. This will enable the initial estimates to be verified or revised.

Predicting food shortage is a technical task, but the outcome is highly political, and it is essential to get the predictions right (within a reasonable margin of error). Famine responses often start too late and go on too long, failing to prevent asset depletion at the start and slowing market recovery at the end. There may be a tendency to 'play safe' and err on the side of predicting shortage. However, it is important to recognize that there are also costs associated with over-predicting shortage. These include costs to maize producers in neighbouring countries and damage to trade networks.

18.6 Can Food Security Be Left to the Market?

It is worth looking at what the cross-border market can and cannot do. The market can:

- Provide a reasonably secure source of maize for those needing and able to buy maize to make up for chronic shortfalls in production.
- By increasing the supply, reduce the domestic price of maize.
- Provide limited buffering in years of poor weather or political disruption, because supplies come from places with slightly different weather patterns and political cycles.

However, the market cannot:

- Provide maize to those households with a shortfall in home production which have insufficient cash to buy maize at times of shortage.
- Supply sufficient maize when there is low production in Malawi and shortages or only moderate surpluses in neighbouring countries, and/or export bans.

In order to deal with the failure of the cross-border market to deliver food security in particular circumstances, a number of interventions are likely to be required, such as welfare transfers, inputs or food aid for the poorest households, judicious releases from Strategic Grain Reserves or timely purchases on the regional or world market.[1]

The evidence suggests that, under current circumstances, cross-border trade cannot alone guarantee food security. However, proper monitoring and prediction of the extent to which the trade will meet Malawi's demand in any given year will provide an opportunity for better management of the intervention options. An intervention approach which better takes into account the cross-border maize trade will be better for food security, better for government finances, better for traders and better for farmers in neighbouring countries.

18.7 Does Starter Pack Harm Farmers in Neighbouring Countries?

Throughout the Starter Pack programme, concerns about negative impact were raised by the Mozambican Government and aid agencies in Mozambique.

At the crudest level, extra maize produced because of Starter Pack probably means less bought from neighbouring-country farmers. But how much less? In the absence of Starter Pack, there is not a one-for-one increase in imports from neighbouring countries. Many Malawi consumers are unable to afford imported maize and go hungry instead.

Accurate figures for loss of market are not available, but qualitative surveys in northern Mozambique report significant hardship caused by periodic failures in the maize market, which are due to a number of factors, one of which is clearly perceived to be Starter Pack (Whiteside, 1998; Arlindo and Tschirley, 2003). The combination of good weather and universal Starter Pack in 1998/99 and 1999/2000 produced a depressing effect for trade volumes and price in northern Mozambique. Whiteside (2002) suggests that about 60% of the reduction in the maize trade in 2000 was due to the universal Starter Pack programme, which was responsible for an average income reduction of at least US$4 per household for the 630,000 northern Mozambican households involved in the border maize trade (a total of US$2.6 million).

Another problem in recent years has been the variation in Starter Pack coverage, as it was scaled down from universal coverage to the Targeted Inputs Programme (TIP) and then scaled up again after the 2001/02 crisis (see Introduction and Chapter 2). Changes in implementation strategy led to a 'boom and bust' impact on markets, which was particularly destabilizing for Mozambican farmers.

Starter Pack may, however, have less negative impacts in the longer term:

- If Starter Pack helps to alleviate and (in conjunction with other initiatives) reduce poverty among Malawian smallholder farmers, this may lead to an overall increase in their purchasing power and demand for products from neighbouring countries.

- A consistent Starter Pack programme might cause a shift in demand away from Mozambican maize. If this trend were consistent, Mozambican farmers could plan for it and move into production of other crops.
- If Starter Pack contributes to macro-economic stability, including exchange rate stability (as suggested in Chapter 15), this should boost regional trade.

As part of the bigger picture, it is important to assess the regional development impacts of Starter Pack in comparison to alternative strategies for addressing Malawi's food security needs (see Chapter 15).[2]

18.8 Comparative Advantage and Regional Development Options

Malawi's neighbours have a comparative advantage in land available for agricultural expansion, including the production of maize. However, as maize is Malawi's main staple food, increasing reliance on imports is a risky strategy unless the country can develop its comparative advantages to earn money to pay for the imports.[3] Malawi's advantages compared with northern Mozambique (Mucavale, 2000) are:

- High population density and labour availability. The country used to be a major labour supplier to South Africa. Labour scarcity is one of the main constraints to increased production in northern Mozambique.
- A relatively more educated and skilled population,[4] with more experience in cash-crop production, particularly tobacco.
- A relatively more developed services and wholesale sector.

This points to at least two potential cross-border development strategies:

1. Concentration on more labour- and land-intensive cash crops in Malawi, with a significant proportion of Malawi's food grown in Mozambique.
2. Malawi, with a more outward-looking service sector and semi-skilled labour, could help to stimulate, and also benefit from, the development of northern Mozambique.

Strong cross-border linkages already exist, particularly between northern Mozambique and southern Malawi:

- Trade goes both ways. While maize flows into Malawi, Mozambique is a significant purchaser of consumer goods, sugar and some food crops, as well as small quantities of Starter Pack seed and fertilizer.
- People migrate temporarily in both directions to work. In particular in years of hunger, Malawians do agricultural *ganyu* in Mozambique.
- People migrate in both directions for more permanent work or to settle. There is some evidence of a net movement of Malawians into Mozambique in recent years because of land availability (Whiteside, 1998, 2002). The Mozambican authorities welcome Malawian 'investors' – those wishing to start large commercial farms – but tend to expel small farmers. Local

populations appear to be relaxed about immigration, as there are usually strong family connections with incomers.
- People travel both directions to access services. In particular, Mozambicans tend to make use of Malawian health facilities, schools, markets and grinding mills.
- Information travels. Border-area Mozambicans listen to Malawian radio broadcasts. A large number of Mozambican households have learnt about agricultural techniques and health issues (e.g. HIV/AIDS) from Malawi.

While Malawi's Starter Pack programme is not incompatible with developing cross-border development strategies, it has:

- Tended to stimulate Malawi's food-crop production more than cash-crop production.
- Hindered development in northern Mozambique by reducing Malawian demand for Mozambican maize.
- (Probably) reduced the need for Malawians to migrate to Mozambique for work.

Thus, there is a need to assess programmes like Starter Pack within a broader sustainable development strategy for the region. National level sustainable development opportunities are limited for Malawi, but by looking beyond national boundaries to the region, more options open up. It is important to ensure that large-scale interventions are compatible with regional development strategies.

18.9 Conclusion

Cross-border trade can make a significant contribution to food security in Malawi. However, evidence from recent years suggests it cannot achieve food security on its own. Interventions are needed to support those households that are unable to afford maize, especially when supplies are short and prices rise. It is vital that Malawi develops the capacity to predict and monitor cross-border trade in order to estimate future deficits and therefore improve the management of interventions. This will require better monitoring and understanding of informal trade. Reducing the barriers to cross-border trade would benefit both neighbouring-country farmers and Malawian consumers.

There is evidence that, in the short term, the Starter Pack programme has had a detrimental effect on neighbouring-country farmers, particularly in Mozambique. This was exacerbated by the changes in programme size from year to year. The regional impact of a longer-term consistent Starter Pack programme may be less negative. However, the key lesson is that any such programme should be assessed, monitored and evaluated as a whole, taking into account the impact on farmer and trader livelihoods in neighbouring countries. From a broader regional development perspective, the future evolution of livelihood security programmes should attempt to reinforce the comparative advantages of Malawi and surrounding areas.

Notes

[1] An issue for consideration is whether purchases should be on the world market or from neighbouring countries. When supply in neighbouring countries is a limiting factor, increasing competition for this supply from institutional purchasers is likely to drive up prices and reduce informal trade.

[2] For instance, a national programme of targeted cash transfers to food-insecure Malawi households would not only have different impacts to those of Starter Pack within Malawi; it would also have different regional impacts.

[3] There are also distributional and gender issues of who has the money to buy food.

[4] Malawi – adult literacy 61%, average adult years of schooling 3.2, primary completion 55%. Mozambique – adult literacy 44%, average adult years of schooling 1.1, primary completion 18% (World Bank, 2004).

References

Arlindo, P. and Tschirley, D. (2003) Regional trade in maize in southern Africa: examining the experience of northern Mozambique and Malawi. Policy Synthesis for cooperating USAID offices.

FEWSNET (2002) Cross border trade during the 2001/02 marketing year in Mbeya and Rukwa Regions.

Macamo, J.L. (1998) Unrecorded cross-border trade between Mozambique and neighbouring countries. A report for USAID.

Minde, I.J. and Nakhumwa, T.O. (1997) Unrecorded cross-border trade between Malawi and neighbouring countries. A report for USAID.

Mozambique Ministry of Industry and Trade/FAO/EC-Food Security Unit (2001) Análise dos custos de transporte na comercialização agrícola em Moçambique.

Mucavale, F. (2000) Analysis of comparative advantage and agricultural trade in Mozambique. USAID Technical Paper 107.

RATES – Regional Agricultural Trade Expansion Support Program (2003a) Maize market assessment and baseline study for Malawi.

RATES – Regional Agricultural Trade Expansion Support Program (2003b) Maize market assessment and baseline study for Zambia.

SIMA (Agricultural Market Information System) Research Team, Mozambique Ministry of Agriculture and Rural Development (2002) Maize exportation: threat to rural food security? Results of a survey of rural households in northern Mozambique Flash 28E. Available at: www.aec.msu.edu/agecon/fs2/mozambique/flash.htm

SIMA (Agricultural Market Information System) Research Team, Mozambique Ministry of Agriculture and Rural Development (2003) Produção e comercialização de culturas alimentares: que expectativas para o presente ano? O ponto de vista dos comerciantes rurais de pequena escala no norte e centro de Moçambique 36P. Available at: www.aec.msu.edu/agecon/fs2/mozambique/flash.htm

Whiteside, M.J. (1998) When the whole is more than the sum of the parts: the effect of cross-border interactions on livelihood security in southern Malawi and northern Mozambique. A report for Oxfam. (Unpublished.)

Whiteside, M.J. (2002) Neighbours in development: livelihood interactions between northern Mozambique and southern Malawi. A report for DFID. (Unpublished.)

Whiteside, M.J. with Chuzo, P., Maro, M., Saiti, D. and Schouten, M-J. (2003) Enhancing the understanding of informal maize imports in Malawi food security. A report for DFID, Malawi. (Unpublished.)

World Bank (2004) World Bank development indicators database. Available at: http://devdata.worldbank.org/external/CPProfile

19 Starter Pack in Rural Development Strategies

ANDREW DORWARD AND JONATHAN KYDD

19.1 Introduction

This chapter examines the potential of Starter Pack to support longer-term pro-poor growth in Malawi. It examines critical constraints to growth in rural Malawi, processes needed for pro-poor growth, and strengths and weaknesses of different strategies and policies for achieving such growth. Discussion focuses on Malawi, but raises issues of wider relevance.

19.2 Critical Constraints to Economic Growth in Rural Malawi[1]

In this section we describe some generic problems facing poor rural areas. We focus on a particular set of problems that increase risks and inhibit productive investments, and suggest possible solutions to these problems. We ignore a number of important social problems that are beyond the scope of this chapter.

A particular characteristic of poor rural areas in Malawi is the lack of a well-developed and diversified monetary economy. In the past, the economy and (directly and indirectly) people's livelihoods were dependent upon two principle activities: agriculture, and migrant labour remittances and returnee savings. The pattern of dependence on agriculture and migrant labour goes back a long way (see, for example, Morton, 1975; Kydd and Christiansen, 1982), but has faced major problems in recent years.

Commercial estate agriculture is in crisis, with very few profitable crops. In the smallholder sector, current stagnation in maize production contrasts with rapidly expanding growth in fertilizer use and hybrid maize production in some areas in the 1980s and early 1990s (Heisey and Smale, 1995; Carr, 1997). While smallholder production of burley tobacco and minor cash crops (such as paprika, birds eye chillies and pigeonpeas) have increased – tobacco was a major beneficiary of the deregulation of the 1990s – these are grown by a minority of farming households (Orr and Mwale, 2001).

Migrant labour opportunities have suffered from long-term declines in opportunities for international migration and from the weakness of commercial agriculture. Diversification into other activities is also difficult, owing to a lack of business opportunities with low capital and skill demands and low risks. Opportunities which do exist, such as petty trading, attract large numbers of people, pushing up supply and – in a generally stagnant economy with static demand – pushing down prices and returns. These problems have been exacerbated by poor macro-economic management and high real interest rates.

Poor rural areas also suffer from weak infrastructure, services and communications (with high costs in physical movement of goods and services in and out of rural areas and high communication costs); low levels of education and literacy, particularly among women; and long-standing problems of poor health, exacerbated by the spread of HIV/AIDS.

The result is a set of very thin markets (markets with low volumes and small numbers of significant participants) for services and goods. These lead to high trading costs and, with poor and costly information services, high risks of transaction failures for buyers and sellers. Traders and producers then need high margins to make their activities worthwhile, but these high margins depress demand.

Market thinness is particularly evident in input, output and financial markets needed for the intensification of maize production, as financial service providers, input suppliers, farmers and output buyers face particularly high costs and risks in investing in these activities, while borrowing is difficult and expensive. Returns on capital, meanwhile, are subject to a range of risks. Maize prices and yields are highly variable, affecting returns to farmers and output traders. Low returns to farmers also pose risks for seasonal finance providers, who face high risks of default even at the best of times. Input stockists face other risks, as inputs such as fertilizer are purchased by farmers in fairly narrow time windows and demand is often very uncertain. If stockists are left with excess inventory, this often cannot be disposed of for another year, carrying high holding costs and often deteriorating in storage.

The result of the problems facing these different players in supply chains is a low-level equilibrium trap and widespread market failure (Dorward *et al.*, 2005). The different players in the supply chain all need high margins to cover their risks and costs, but the total margin in the supply chain cannot accommodate all these margins. Significant investments are therefore stifled or limited to low-cost, small-scale and marginal or opportunistic activities rather than the large-scale, committed enterprises needed to foster significant growth. This is not to say that most rural Malawians are not deeply embedded in some markets: for instance maize and labour markets (*ganyu*) are critical to the livelihoods of many poor households (see Chapter 9). Rather, these markets tend to involve very small and local players and transactions, and the specific supply chains needed for rural people to intensify farm production or to start adequately capitalized non-farm enterprises tend to be absent or very weak.

19.3 Short- and Long-term Policies for Food Security, Poverty Reduction and Economic Growth

Where markets are thin and not working properly, policies cannot rely on markets to coordinate and deliver services. Market-based approaches to food security do not work in poor rural economies, as Malawi's 2001/02 crisis demonstrated, and such policies cannot be effective without prior development of markets and of firms within them. Market-based poverty-reduction polices face the same problem. There is therefore a need for an approach to immediate food security, medium term poverty reduction and longer-term pro-poor growth policies that distinguishes between:

- the short-term need for all policies to work in the absence of effective markets or market economy organizations;
- the medium-term need for development of an effective market economy; and
- eventual reliance of policy interventions on markets and firms in such an economy.

Policy makers must construct consistent sets of policies that complement each other in pursuing both short- and long-term objectives (immediate welfare improvements for the vulnerable and pro-poor growth respectively) and in their immediate and eventual policy instruments (non-market and market economy based, respectively). The aim should be a policy set which provides consistency and complementarity of policies across different policy goals and time periods. Policies should also allow for and address the context in which they must operate – not only the lack of market development in the rural economy, discussed above, but also the historical context (affecting institutions and people's expectations and behaviour) and opportunities and constraints arising from governance, resources, infrastructure, health and education services and status, HIV/AIDS, gender relations, the environment and current activities in the rural economy.

The distinction we draw between immediate non-market policy interventions and longer-term market development and reliance policies is illustrated in Table 19.1. This provides a helpful framework for examining the contributions of alternative policy interventions supporting food security, poverty reduction and growth in poor rural areas. The analysis is open to the criticism that it is 'pre-Sen' in its emphasis on short-term food production increases. We accept the logic of Sen's general approach to famines and food security (Sen, 1981), but the standard entitlements analysis assumes that markets work reasonably well, whereas in Malawi this is not the case.[2]

Before we turn to specific agricultural policy interventions in Malawi, including Starter Pack, it is necessary to examine the context of (i) global experience of large-scale success in improving food security, reducing

Table 19.1. Policy requirements for short- and long-term achievements of food security, poverty reduction and economic growth.

Policy Goals	Requirements for Short-/Medium-term Achievement (Policy Purpose)	Requirements for Medium-/Long-term Achievement (Policy Purpose)
Food security: Secure and affordable access to food	Increased food self-sufficiency (household and national) with **food delivery** and/or **productivity-enhancing** safety nets and humanitarian response	Increased household and national food market access (low and stable cost, secure, timely) through wider entitlements with (mainly) **market-economy-based** safety nets and humanitarian response
Poverty reduction: Real incomes of the poor increased and more secure, through low food costs, higher returns to labour, and safety nets	Safety nets to increase/secure real incomes and develop/protect assets (see above)	Broad-based growth with opportunities and wages for unskilled rural labour, low food prices, and safety nets and humanitarian response as above
Rural economic growth: Increased levels of local economic activity, with stable income opportunities supporting poverty reduction and food security	Short-/medium-term achievement not possible	Macro-economic stability and low interest rates; growth in agricultural and non-agricultural sectors tightening labour markets and raising real incomes with stable/affordable food prices. Development of market economy. Initial growth must be achieved without depending on (non-existent) markets or firms

Source: Modified from Dorward and Kydd (2003).

poverty, and achieving economic growth and (ii) the role of agriculture in achieving pro-poor economic growth in rural areas.

19.4 Global Experience of Growth and Welfare Policy Development in Poor Rural Areas

In the early 1960s there were grave concerns about poverty and food security in Asia, with predictions of mass food shortages as populations grew rapidly and food production remained stagnant. There was particular concern about India, Bangladesh and China, which had all experienced severe famines within the previous 20 years. Now, however, the region has achieved remarkable success in improving food security, reducing poverty, and promoting economic growth, although there remain significant numbers of poor and food insecure people in parts of (particularly South) Asia.

Dorward *et al.* (2004b) review policies in Asia and other parts of the world where rapid and large-scale agricultural transformations have occurred in poor rural areas and driven rapid economic growth and poverty reduction. They suggest that these successful agricultural transformations have involved three phases, with different state policies promoting institutional and technical change: first, a period of basic investments to establish suitable conditions for widespread adoption of intensive cereal technologies; second, a 'kick-starting markets' phase, when government interventions supported coordinated service delivery to farmers to enable wider access to seasonal finance and input and output markets at low cost and low risk; and finally, withdrawal of the state from market intervention once a broad-based agricultural transformation had been achieved. The final phase has been characterized by increased traded volumes lowering unit transaction costs in credit, savings, inputs and produce markets, and growing volumes of non-farm activity. Such policies did not always succeed, and indeed there are many countries (particularly in Africa) where they failed dramatically, causing long-term economic damage. However, where successful they provide the dominant, perhaps only, pattern of widespread and rapid pro-poor growth in poor rural areas.

The three phases of basic investment, kick starting markets and withdrawal of the state represent, with regard to policies for pro-poor growth, movement through the three policy phases discussed in the previous section: (i) initial reliance on non-market interventions, (ii) development of a market-based economy and (iii) reliance on a market-based economy. A similar evolution of welfare policy is also apparent in many of these countries, as government grain distribution systems (in Bangladesh for example) have been scaled down or withdrawn and greater emphasis put on income support, with reliance on markets to link this to food security.

19.5 Pro-poor Growth in Rural Areas: the Role of Agriculture

Three broad strands of theory and evidence suggest a major role for agriculture in driving economic growth in poor rural areas: sectoral growth theories, multiplier analysis and livelihoods analysis.

Sectoral growth theories argue that in poorer countries where the agricultural sector accounts for a large proportion of Gross Domestic Product (GDP) and an even larger proportion of employment, a dynamic agricultural sector can make five contributions to broader development: increasing agricultural productivity for capital investment in agriculture and for the steady release of surplus capital and labour to other sectors of the economy; providing the major source of export earnings and of food; keeping food prices down; and generating domestic income to stimulate demand for local goods and services

(Mellor, 1986; Timmer, 1988). Arguments that agriculture drives pro-poor growth in poor rural areas are strongly supported by econometric studies of sectoral indicators and economic growth and poverty reduction in the latter part of the 20th century (Kydd *et al.*, 2004).[3]

A second set of arguments, based on analysis of multipliers and linkages in rural economies, identifies a critical role for agriculture in driving growth in many poor rural areas, but also recognizes the role of non-agricultural activities in supporting growth and distributing its benefits within rural communities (Dorward *et al.*, 2003; Kydd *et al.*, 2004). This analysis focuses on two mechanisms which can drive growth in an economy: growth in production of tradables (leading to direct increases in local employment and producer incomes) and in production of non-tradables on which people spend a large share of their income (where increasing local production depresses prices and thus increases real consumer incomes).[4]

In poorer, more remote areas there are unlikely to be many tradable non-farm activities, apart from mining, that offer broadly based employment opportunities. Only as communications with urban areas improve do opportunities for other non-farm tradable activities develop, but trading links and remittances are generally more important than local employment in generating non-farm tradable activities (Bryceson, 1999). Non-farm activities that do exist tend to have high barriers to entry (Reardon, 1998). Farm activities therefore generally have more potential to drive pro-poor growth through expansion in cash-crop or food-crop production. Non-tradable non-farm activities (together with livestock and horticultural production) are important in supporting growth by responding to increased demand for non-tradables with high marginal budget shares.

The final set of arguments for an important (but not exclusive) role for agriculture in pro-poor growth in rural areas comes from an analysis of the structure of poor people's livelihoods. These arguments highlight complex, interrelated and changing roles of agriculture and own-farm and non-own-farm activities in poor people's livelihoods.

Based on detailed livelihood and rural economy models in Malawi, Dorward (2003) puts forward arguments highlighting the complementary roles of farm and non-farm growth. While the non-farm sector provides a major part of rural incomes and often supports farm investment, he concludes that development of the farm sector should initially be the highest priority because:

1. Food prices and production are critically important to real incomes of the poor (and without reliable integration with wider markets, local production may have a critical impact on food prices and access).
2. Own-farm labour use has a major impact on local labour markets (if it were to contract, this would release a flood of labour into the local labour market, depressing wages in all activities).
3. Poor households' access to casual employment and wages, critical to their livelihoods, depends in part upon on-farm labour demand among less-poor farmers.

4. Own-farm labour can have higher average and marginal returns to labour than off-farm employment at critical times of year.
5. Non-farm growth opportunities depend largely upon agriculture driving the local economy, as in most rural areas there are relatively few opportunities for the non-farm sector to drive growth.

Own-farm smallholder production, and its continued development, are therefore critically important to the poor, but so is the non-farm sector. Both must develop together so that if the smallholder agriculture sector grows then the non-farm sector can, with time and improved markets, increasingly support and take over from agriculture. However, the immediate challenge is to get Malawian smallholder agriculture to provide rural Malawians with secure access to food, and to secure and higher incomes in a growing economy.

19.6 Policy Interventions Promoting Pro-poor Growth: a Review

We are now in a position to examine the specific contributions of different types of policy intervention to food security, poverty reduction and pro-poor growth. Of these three policy goals, the last is perhaps the most challenging, but without economic growth the costs of food-security and poverty-reduction interventions will continue to increase. Rural welfare and development policies should therefore start from a set of policies that will stimulate growth, and complement them with welfare policies that support, rather than undermine, such growth.

How can growth be achieved and the problems of thin markets overcome? We identify six broad and related changes needed for growth:

1. *Development of coordination mechanisms and of relationships of trust* between different players in the supply chain.
2. *Large-scale investments to 'kick-start' activities* and get volumes up beyond the critical point at which margins fall sufficiently, together with infrastructure investments to reduce transport and communications costs and improve access to markets and services.
3. *Insurance mechanisms* that protect players from some of the natural, market or transaction risks they face.
4. *Market interventions* to modify margins at critical points in the supply chain.
5. *Technical change* that reduces the need for purchased inputs and seasonal finance.
6. *Economic growth* that increases volumes of economic and market activity more generally in the local economy.

This section examines the principal development policy interventions employed or considered in the past and present in Malawi. Our focus is on the first four needs for change listed above and on the farm sector, in particular maize. Although development of cash crops is also very important, indeed complementary,[5] growth in intensive cereal production may offer better linkages and hence more potential for broad-based pro-poor growth (Dorward and Morrison, 2000).

19.6.1 Maize price stabilization

Where prices vary widely and unpredictably between seasons (as is the case in Malawi), stabilization of maize prices can offer substantial benefits to both producers and consumers. With a guaranteed minimum price, producers gain from protection against sudden price falls and, provided that the price is high enough, this should make them more willing to invest in surplus maize production. Maize traders may also benefit, although they lose the opportunity to make large profits from holding stocks in times of scarcity. Consumers buying maize, on the other hand, gain from a guaranteed maximum price. This may allow them to reduce own production of maize and specialize in other more profitable activities without fear of food insecurity when prices rise (Dorward, 1999; Orr and Orr, 2002).

We have used farm household models with 1997/98 prices to investigate the effects of price stabilization.[6] Preliminary results suggest that if maize price uncertainty were removed, this would lead to reductions in equilibrium maize prices but increases in aggregate maize production, with more specialization by some (surplus producer) households and some movement by other households out of maize production into other crops and non-farm activities. This would be accompanied by modest improvements in real incomes for the poor, but small losses in real incomes for other households.

However, the way price stabilization is implemented is critical to its impact. Guaranteeing minimum prices requires the ability to intervene in the market and buy when prices are falling. Surpluses then have to be either stored in some form of national grain reserve or sold (probably at a loss). Conversely, guaranteeing maximum prices requires sales into the market during periods of scarcity, and these sales have to be sourced either from imports (probably at a loss) or from national stocks. Government interventions to stabilize prices are therefore costly (see Chapter 15); they are frequently hijacked for political and patronage purposes, and a common lack of transparency in their operation imposes very high risks on private traders – who then withdraw from the market. Recent experience with the management of buffer stocks in Malawi has been expensive and ineffective (Whiteside, 2003). In the 1970s and early 1980s, however, this approach had a much more positive record: we estimate that price stabilization (and price setting) by the Agricultural Development and Marketing Corporation (ADMARC) required relatively small subsidies as it was able to make considerable profits through arbitrage across intra-seasonal maize price differences.

There are two further difficulties with price stabilization policies. First, porous borders of small landlocked countries like Malawi make it very costly to defend a price within national borders without also effectively undertaking the much larger task of modifying prices in neighbouring countries (see

Chapter 18). Second, it is very difficult to stabilize prices without at the same time 'setting' them above or below their long-term equilibrium, because it is difficult to determine equilibrium prices once the state starts intervening, and there are normally strong political pressures to use the price stabilization mechanism to change average price levels.

Poulton and Dorward (2003) review these difficulties but conclude that in countries such as Malawi 'there remains a case for a degree of state intervention in staple food markets that goes beyond the minimal contingency stock to protect against delays in private importation.'

19.6.2 Maize price support

If maize prices are to be stabilized, a critical question has to be faced regarding stabilized price levels. The arguments for low and high maize prices are finely balanced.

It has long been recognized that the majority of Malawi's poor rural people are deficit maize producers whose ability to access food is damaged by high maize prices. The Starter Pack/Targeted Inputs Programme (TIP) evaluation documented the magnitude of this problem in the early 2000s (see Chapter 8). The extent of a positive supply response by smallholders is also questioned, as poorer farmers may find that higher maize prices prior to harvest lead to higher maize expenditures and, due to credit constraints, this may actually reduce the resources available for maize production. Cultural factors and risk aversion also make most Malawian households try to grow as much of their own household maize requirements as possible, irrespective of market prices (see Chapter 9). Since the number of households producing maize to sell is relatively small, benefits from higher prices may be largely irrelevant to national maize production. On the other hand, stable low maize prices may encourage diversification away from maize production into other more remunerative activities (Dorward, 1999; Orr and Orr, 2002).

While the short-term costs of high maize prices for poor deficit households are indisputable, other impacts of higher maize prices are less well understood. Although the negative effects of higher maize prices on maize production by the poor have been widely reported (UNICEF, 1993; Pearce *et al.*, 1996) and modelled (Dorward, 1996, 1999, 2003), their extent, particularly in southern Malawi, has never been properly investigated. Furthermore, while the number of significant surplus maize producers may be small compared to the total number of rural households, farm household models constructed by the authors suggest that they nevertheless account for a substantial proportion of total maize purchases in Malawi. Dorward *et al.* (2004c) also find that such producers show a significant positive supply response to increasing maize prices over a limited range of prices, and benefit from these higher prices. Poorer households, however, suffer from higher prices and, over higher price ranges, show a flat, or in some cases negative, supply response.

The situation becomes more complex when impacts of increasing maize prices on rural wages are allowed for, taking account of: (i) consumption multipliers from income gains and losses by net maize surplus and deficit households; and (ii) increases in labour demand in more intensive maize production.[7] Less poor households gain from increasing maize prices as they are able to increase production and become surplus producers, while poor, maize-deficit households lose. However, partial equilibrium models suggest that over higher ranges of prices larger maize price increases can push up wage rates, and these can substantially moderate the negative effects of maize price increases on poor households' real incomes. This finding is, however, sensitive to assumptions about labour market behaviour, about which there is very little information.[8]

An important point that emerges from this discussion is the sensitivity of impacts of increasing maize prices (i) to the range over which these prices occur and (ii) to other policy interventions. This arises because (i) poorer households' responses to increasing maize prices are increasingly constrained at higher maize prices and (ii) there are threshold prices above which purchased inputs and more intensive maize production by less poor households become profitable, for subsistence and for sale. Increasing prices above these thresholds may deliver only limited supply responses, while price increases across these thresholds may deliver very high supply responses, with associated increases in labour demand providing benefits to the (able) poor. The willingness and ability of less-poor households to intensify maize production in response to increased prices will also vary, however, with access to seasonal finance for input purchases. This needs action to address coordination problems in the supply chain, particularly those facing input stockists and suppliers of seasonal finance.

Two further issues need to be considered in any discussion of maize price support: (i) costs of dealing with maize surpluses from increased domestic production and imports; and (ii) decreased food security and welfare if the poor face higher maize prices. Price support interventions must be designed and implemented to limit imports, while prices should be set to avoid regular domestic surpluses.[9] Some form of targeted maize- or cash-distribution system to improve food access for the poor should also accompany policy interventions raising maize prices.

19.6.3 Input subsidies

While some argue that Starter Pack represents an input subsidy, we analyse it as a transfer (see Section 19.7). Here, we consider policies that subsidize input purchases by farmers.

The advantages and disadvantages of input subsidies in stimulating more intensive agricultural production have been widely discussed (for example Ellis, 1992). Proponents argue that input subsidies can help overcome critical informational, price and transaction cost distortions in poor rural areas to generate both efficiency and equity gains in the early stages of an agricultural transformation, citing Indonesia as an example of successful use of input subsidies. Critics argue that there are large leakages to existing input users and (in landlocked countries) across borders, that they promote inefficiency through wastage and over-use, that they represent a large fiscal burden, that they offer benefits primarily to the less poor, and that unless they are very large they do not address the seasonal finance constraint to purchased input use by poorer farmers.

How input subsidies are implemented is critical to their effectiveness and efficiency. Some criticisms of input subsidies (for example high leakages) may be partially addressed by careful design of delivery mechanisms, although such arrangements carry large administrative challenges and can stimulate corrupt practices. Input subsidy mechanisms can also promote the capacity of input stockists to supply inputs to farmers in poor rural areas (by stimulating demand and reducing coordination risks) or, alternatively, crowd them out and increase their risks – depending on the way that they are implemented.

Empirical evidence on the impact of input subsidies in Malawi is limited, and impacts depend on relative input, maize and other crop prices as well as on subsidy mechanisms. Dorward *et al*. (2004a) report findings from household, rural economy and Computable General Equilibrium (CGE) modelling of 20% subsidies on fertilizer purchases in Malawi. Both partial and general equilibrium models find that although the subsidies go to less poor farmers, the poor benefit from impacts on maize prices and rural wages. Contrary to expectations, the fertilizer subsidies turn out to be pro-poor in the sense that they not only increase incomes of the poor, but increase them more (in proportionate terms) than the incomes of the less poor. The CGE model also estimates that such subsidies promote agricultural growth.

19.6.4 Interlocking arrangements

Interventions in output and input markets do not normally address the problems of coordination failure in supply chains.[10] Interlocking arrangements exist to address these problems, describing mechanisms where farmers' purchases of inputs, seasonal borrowing and produce sales are integrated in directly related transactions. Poulton *et al*. (1998) define interlocking as: 'Provision of seasonal inputs on credit using the borrower's expected harvest of the crop in question as a collateral substitute to guarantee loan repayment.' Interlocking contracts underpin most successful smallholder seasonal finance programmes in Africa. Two very successful Malawian examples are the old Smallholder Agricultural Credit Administration (SACA) system for financing maize inputs supplied by ADMARC (see Chapter 10) and the current tobacco credit system, with inputs supplied by local input dealers who are paid by the Malawi Rural Finance Company (MRFC) with loan recovery through sales at the auction floor.

The conditions for interlocking systems to work are quite restrictive, requiring a profitable crop and some form of monopsonistic, or at least oligopolistic, market (Dorward *et al.*, 1998). These conditions are more common with cash crops and rarely arise with food crops. The success of Malawi's SACA programme was unique and dependent upon: effective operation of ADMARC as a monopsonistic buyer; effective and close relations between SACA, the extension services and ADMARC with well-functioning farmer credit groups; maize price stabilization and careful setting of maize and input prices; and political commitment to enforcing credit repayments, with defaulters being denied further access to credit and inputs and, *in extremis*, suffering confiscation of assets.[11]

Where they can be made to work effectively, interlocking arrangements offer a number of benefits to the different players in the supply chain. They can dramatically reduce coordination and transaction risks, and at the same time offer economies of scope across the different transactions (finance, inputs and outputs). By tying the different transactions together within a personalized service provider–farmer relationship, they also allow specific subsidies to be offered within these relationships, and cross-subsidization across financial, input and output transactions. Thus, one way of targeting maize price support and limiting its fiscal costs is to offer higher prices only to farmers buying inputs on credit, with these prices offered for only a specific quota per farmer. This could also be considered a form of input or credit subsidy – and could limit (but not eliminate) leakages. Elements of this model (tying together output quotas with input packages from government agencies) have been central to Chinese agriculture's transition to more market-based transactions.

A major challenge exists in establishing food-crop interlocking systems that benefit farmers, contain incentives that promote efficient operations by service providers, are effective in enforcing credit repayment and foster longer-term development of private-sector activity in service delivery. There are many examples of effective cash-crop interlocking systems operating within the private sector, but their operation for food crops requires at least market regulation by the state.

19.6.5 Farmer groups

Farmer groups offer another way of improving coordination and reducing coordination and transaction costs and risks along the supply chain. This is achieved at the price of significant coordination costs within the farmer organization, and therefore tends to be most effective where accessing services for cash-crop production offers substantial benefits to producers. The Malawian SACA experience with maize can be interpreted as a situation where agricultural policy made maize into a profitable cash crop for members of farmer clubs. There are some highly successful farmer organizations in Malawi today, but almost all of them are organized around production of a cash crop.

19.7 Transfers and Pro-poor Growth

Transfers do not constitute pro-poor growth strategies. They fall into the category of immediate non-market policy interventions to achieve food security and/or poverty reduction (see Table 19.1). However, it is important that they do not impede critical longer-term growth and development processes.

19.7.1 Food, cash and input transfers

Transfers usually involve cash, food or inputs which are targeted at poorer people or universally provided. We do not attempt to review all the different types of transfer schemes, but it is important to note the following points.

First, food and cash transfers provide immediate benefits to recipients, whereas input transfers provide resources for future production, unless they are immediately sold on. As poor households' labour and other investments in future production may be constrained by the need to devote these resources to meet current consumption needs, cash and food transfers may also offer significant benefits for future production. Thus household, partial and general equilibrium model simulations reported by Dorward (2003) and Dorward *et al.* (2004c) estimate that pre-harvest universal and targeted cash transfers raise both the welfare and the agricultural productivity of the poor.

Second, transfers are likely to affect local prices. Thus, cash transfers tend to inflate local prices, grain transfers to depress grain prices, and input transfers to depress purchased input demand, if not prices (although this was very limited in the case of Starter Pack – see Chapter 10). The extent of price changes (and wage changes, as labour markets may also be affected) depends upon transfer volumes in an area and elasticities of supply and demand for affected commodities. However, food and input transfers can be designed and managed to involve private-sector players, using large welfare investments to 'kick-start' private-sector capacity through the provision of large and secure markets.

Third, targeted transfers are difficult to manage. Self-targeting through low-paid work may help construct roads or other public infrastructure, but those unable to work are excluded, and in very poor communities wages may be unacceptably low if they exclude the relatively less poor. Administrative and community targeting can be very divisive, is prone to abuse and frequently involves large leakages to less poor people and omission and/or stigmatizing of the poor (see Chapter 3 and Chapter 11).

Transfers therefore differ in the immediacy of their welfare impacts, their impacts on markets and market development, and the effectiveness of targeting. They also differ in their longer-term development impacts on productivity and market development.

19.7.2 Starter Pack

Starter Pack is a universal inputs transfer. Here we consider whether it contributes to growth and development or undermines these longer-term processes. We focus on the major components of the Starter Pack: maize seed and fertilizer.

There seems to be little doubt that universal Starter Pack increases productivity and output of maize (see Chapter 8). However, the evaluation surveys could not – and did not attempt to – separate out: (i) direct productivity effects; (ii) the effects of households re-organizing their activities to get more advantage from their Starter Pack; and (iii) impacts on food prices and wages (which themselves affect farmers' cropping activities). Dorward (2003), using 1997/98 prices but taking account of wage and maize price changes (with admittedly arbitrary assumptions about wider maize and labour market elasticities), estimates that Starter Pack led to an increase in real incomes of 9% for poor households, and an 11% increase in maize production. Without wage and maize price changes, however, poor household real incomes are estimated to increase by only 6%.

As wage and maize price changes appear to play a large role in determining the benefits of Starter Pack to the poor, reducing its scale and targeting it at poorer households leads to lower benefits for those households, as gains from cheaper maize purchases or higher wages are considerably reduced. Dorward (2003) estimates that targeting Starter Pack at the poorest 50% of the population leads to only 5% increases in real incomes for poorer households receiving the pack.

Thus, in terms of the scheme presented in Table 19.1, universal Starter Pack should make a significant contribution to food security and poverty reduction – benefits that would be much more limited in scale and scope with a targeted free inputs programme. However, the relatively small real income gains from Starter Pack do not provide enough of a stimulus to drive forward a process of growth. Indeed, by depressing maize prices, it may undermine important growth contributions from less poor households that engage in more intensive labour-demanding maize production. There are also questions about the ability of labour-constrained households to take advantage of free inputs, and about the extent to which Starter Pack can support market development – either undermining or building up input stockist systems depending upon the way that inputs are sourced and distributed.

19.8 Conclusion and Recommendations

This chapter has argued that maize production is central to any pro-poor growth processes in Malawi (although it is not suggested that increasing maize production is either the only process needed for pro-poor growth, or a sufficient one to drive such growth on its own). Particular attention needs to be paid to the coordination problems constraining the market and service development necessary to increase land and labour productivity in maize cultivation. A distinction is then made between, on the one hand, more immediate policies in poor rural areas (which cannot rely on absent coordinated markets and firms), and, on the other hand, policies that are more reliant on a market-based economy but which will only be appropriate once markets and firms have developed. As a result, while short-term policy objectives may emphasize transfers to provide immediate improvements to the welfare of poor people, an important medium- to long-term policy objective must be to promote market and organization development in a way that overcomes current coordination failures. It then becomes critical that the implementation of immediate, short-term welfare improvement policies support rather than undermine the long-term goal of developing larger markets and firms in a market-based economy in rural areas.

The analysis identified the contributions of different policy interventions in Malawi to pro-poor growth. Different interventions have different strengths and weaknesses, and much depends upon the way that they are designed and implemented. A combination of different interventions is needed, with design and implementation complementing each other in promoting immediate welfare, development of markets and firms, and pro-poor growth.

Universal Starter Pack is unlikely to feature in such a set of policies, as it is too blunt an instrument and one of its main benefits to the poor – lower maize prices – may actually impede critical development processes. Targeted input transfers, on the other hand, are unlikely to reduce maize prices and to push up *ganyu* opportunities and wages, and will therefore only help the able poor who can wait until after harvest for their benefits: there will be limited benefits to those who cannot use the inputs or who desperately need food or cash for immediate consumption (these can only benefit by selling the inputs). Nevertheless, targeted input transfers may well have a role to play in an integrated set of interventions promoting both immediate welfare gains and longer-term pro-poor growth through the development of markets and firms.

Such a programme might involve the following elements:

- Price stabilization policies that provide a minimum floor price for maize during harvest and the immediate postharvest periods, and then a maximum price ceiling at other times of the year. Price stability would encourage more specialization among producers and consumers, and thus deliver development benefits. It would also offer welfare benefits to people who sell maize at very low prices at harvest time to meet immediate cash needs, only to buy maize back later in the year at higher prices.
- Further welfare benefits for the poor could be provided through transfers in the form of maize handouts to the destitute and self-targeting inputs for work for the able poor (the latter would also contribute to the development of rural infrastructure).
- More intensive maize production with the development of a coordinated supply chain could be stimulated through an interlocking system where farmer group members can obtain inputs on credit repayable in maize valued at above market prices – with a further quota of above market price cash sales. Arrangements could be set up for loans, input sales and produce

purchases to be handled by private firms, with a district-level coordination system identifying defaulters. A maximum period could be set for individual farmers' access to interlocked loans on preferential terms, to limit the scale and time over which subsidies are offered. Some insurance against weather-related crop failure would also need to be established.

Such a programme would pose substantial design and implementation challenges to avoid inefficiency and corruption. It would need to be costed (with costs compared to costs of relief and welfare programmes needed to respond to chronic and acute food insecurity), and would require a long-term commitment from donors and government to see it through. Finally, it should complement other growth initiatives within agriculture and in the non-agricultural sector.

Notes

[1] This section draws heavily on material published in Dorward and Kydd (2004).

[2] See Poulton and Dorward (2003) for further elaboration of these arguments.

[3] Further arguments from the 'New Economic Geography' suggest that more remote rural areas suffer from disadvantages in accessing information and markets. As a result, urban areas have a strong comparative advantage for many economic activities: in 'the deep countryside' only activities with a strong natural resource base, local processing of agricultural products, and non-tradable services for the rural population will survive (Wiggins, 2001). Exceptions to this occur where labour costs are much lower in rural (as compared with urban) areas, and there are good communications and road networks in rural areas. These latter conditions do not hold in Malawi and labour-intensive industries often struggle even in urban areas.

[4] Tradable goods and services are those that may be imported or exported to or from the area. Although the term is often associated with international trade, it is equally applicable to *intra*national trade between different districts or between rural and urban areas.

[5] Intensification of food- and cash-crop production should be seen as complements, and more intensive cash-crop production should support rather than compete with maize production. For instance, tobacco currently plays a major role in Malawi's rural economy. Nevertheless, the search for a wider range of cash crops and for greater smallholder opportunities in cash-crop production continues to present a challenge. This limits the extent to which cash-crop production can be relied upon to drive pro-poor growth.

[6] Analysis used a modified version of the models described in Dorward (2003). Results are sensitive to assumptions about producers' and consumers' risk aversion and about wider labour and maize market demand and supply elasticities.

[7] Conclusions here depend upon assumptions about the elasticities of labour demand outside the informal rural economy. Using a partial equilibrium model, Dorward et al. (2004c) estimate elasticities of maize production to vary from 0.1 to 3, over different maize price ranges. Using a CGE model, elasticities are estimated as varying between −1 and +1, although the larger negative elasticity estimates need to be treated with some caution.

[8] The models also do not allow for constrained access to input supplies and probably under-estimate seasonal finance constraints inhibiting households' fertilizer purchases. Responses to price increases will therefore probably be reduced unless accompanied by measures to address seasonal finance and input supply constraints. Dorward et al. (2004a) also show that reducing input and output marketing margins could bring substantial growth and poverty-reduction benefits.

[9] The risks and costs of a substantial surplus occurring in the occasional 'good year' (requiring exports at below domestic purchase price) should, however, be offset against the relief costs that would otherwise be incurred in importing maize in deficit years.

[10] Economic coordination can be defined (Poulton *et al.*, 2004) as 'efforts or measures designed to make players within a market system act in a common or complementary way or towards a common goal'. Coordination failure is thus a failure to make an investment (e.g. buy and apply fertilizer) due to: (i) the possible absence of complementary investments by other players at different stages in the supply chain or (ii) risks of opportunistic behaviour by other players, e.g. strategic loan default by farmers rendering credit supply unprofitable (Dorward and Kydd, 2004). Coordination failure can arise even in supply chains which have the potential to be profitable given the available price margins. It can be reduced by private, collective or government action or a mixture of all three.

[11] The relative importance of confiscation of assets in ensuring loan repayment is a matter of debate. Experience around the world with successful micro-finance programmes and with interlocking systems in cash-crop production suggest that group pressure and denial of access to future borrowing opportunities can provide very effective incentives for repayment.

References

Bryceson, D. (1999) Sub-Saharan Africa betwixt and between. Working papers. African Studies Centre, University of Leiden, Netherlands.

Carr, S.J. (1997) A green revolution frustrated: lessons from the Malawi experience. *African Crop Science Journal* 5(1), 93–98.

Dorward, A.R. (1996) Modelling diversity, change and uncertainty in peasant agriculture in northern Malawi. *Agricultural Systems* 51(4), 469–486.

Dorward, A.R. (1999) Farm size and productivity in Malawian smallholder agriculture. *Journal of Development Studies* 35(5), 141–161.

Dorward, A.R. (2003) Modelling poor farm-household livelihoods in Malawi: lessons for pro-poor policy. Centre for Development and Poverty Reduction, Department of Agricultural Sciences, Imperial College, London.

Dorward, A.R. and Kydd, J.G. (2003) Work in progress: policy analysis for food security, poverty reduction, and rural growth in Malawi. Centre for Development and Poverty Reduction, Department of Agricultural Sciences, Imperial College, London.

Dorward, A.R. and Kydd, J.G. (2004) The Malawi 2002 food crisis: the rural development challenge. *Journal of Modern Africa Studies* 42(3), 343–361.

Dorward, A.R. and Morrison, J.A. (2000) The agricultural development experience of the past 30 years: lessons for LDCs. Imperial College at Wye for FAO.

Dorward, A.R., Kydd, J.G. and Poulton, C. (1998) Conclusions: NIE, policy debates and the research agenda. In: Dorward, A., Kydd, J. and Poulton, C. (eds) *Smallholder Cash Crop Production under Market Liberalisation: A New Institutional Economics Perspective*. CAB International, Wallingford, UK.

Dorward, A.R., Poole, N., Morrison, J.A., Kydd, J.G. and Urey, I. (2003) Markets, institutions and technology: missing links in livelihoods analysis. *Development Policy Review* 21(3), 319–332.

Dorward, A.R., Kydd, J.G., Morrison, J.A. and Poulton, C. (2005) Institutions, markets and economic co-ordination: linking development policy to theory and praxis. *Development and Change* 36(1), 1–25.

Dorward, A.R., Wobst, P., Lofgren, H., Tchale, H. and Morrison, J.A. (2004a) Modelling pro-poor agricultural growth strategies in Malawi: lessons for policy and analysis. Centre for Development and Poverty Reduction, Department of Agricultural Sciences, Imperial College, London.

Dorward, A.R., Kydd, J.G., Morrison, J.A. and Urey, I. (2004b) A policy agenda for pro-poor agricultural growth. *World Development* 32(1), 73–89.

Dorward, A.R., Poulton, C., Tchale, H. and Wobst, P. (2004c) Disaggregated impacts of agricultural policy reform on poor rural households: linking household, rural economy and economy-wide analysis. Paper prepared for the OECD Global Forum on Agriculture: Designing and Implementing Pro-Poor Agricultural Policies. Wye, Imperial College, London.

Ellis, F. (1992) *Agricultural Policies in Developing Countries*. Cambridge University Press, Cambridge, UK.

Heisey, P.W. and Smale, M. (1995) Maize technology in Malawi: a green revolution in the making? *CIMMYT Research Report* No. 4. CIMMYT, Mexico DF.

Kydd, J.G. and Christiansen, R. (1982) Structural change in Malawi since independence: consequences of a development strategy based on large scale agriculture. *World Development* 10, 355–375.

Kydd, J.G., Dorward, A.R., Morrison, J.A. and Cadisch, G. (2004) Agricultural development and pro poor economic growth in sub Saharan Africa: potential and policy. *Oxford Development Studies* 32(1), 37–57.

Mellor, J.W. (1986) Agriculture on the road to industrialisation. In: Lewis, J.P. and Kallab, V. (eds) *Development Strategies Reconsidered*. Transaction Books, New Brunswick.

Morton, K. (1975) *Aid and Dependence: British Aid to Malawi*. Overseas Development Institute/Croom Helm, London.

Orr, A. and Mwale, B. (2001) Adapting to adjustment: smallholder livelihood strategies in Southern Malawi. *World Development* 29(8), 1325–1343.

Orr, A. and Orr, S. (2002) Agriculture and micro enterprise in Malawi's rural south. *Agren Network Paper* 119, Overseas Development Institute, London.

Pearce, J., Ngwira, a. and Chimseu, G. (1996) Living on the edge: a study of the rural food economy in the Mchinji and Salima districts of Malawi. Save the Children Fund – UK, Lilongwe, Malawi.

Poulton, C. and Dorward, A. (2003) The role of market based economic development in strengthening food security. Paper prepared for the ODI Southern Africa Forum on Food Security. Overseas Development Institute, London.

Poulton, C., Dorward, A., Kydd, J., Poole, N. and Smith, L. (1998) A new institutional perspective on current policy debates. In: Dorward, A., Kydd, J. and Poulton, C. (eds) *Smallholder Cash Crop Production under Market Liberalisation: A New Institutional Economics Perspective*. CAB International, Wallingford, UK.

Poulton, C., Gibbon, P., Hanyani-Mlambo, B., Kydd, J., Maro, W., Larsen, M.N., Osorio, A., Tschirley, D. and Zulu, B. (2004) Competition and coordination in liberalized African cotton market systems. *World Development* 32(3), 519–536.

Reardon, T. (1998) Rural non-farm income in developing countries. In: *The State of Food and Agriculture 1998*. FAO, Rome.

Sen, A.K. (1981) *Poverty and Famines: An Essay on Entitlement and Deprivation*. Clarendon, Oxford, UK.

Timmer, C.P. (1988) The agricultural transformation. In: Chenery, H. and Srinivasan, T.N. (eds) *The Handbook of Development Economics*, Vol. I. Elsevier Science, Amsterdam.

UNICEF (1993) Malawi : situation analysis of poverty. United Nations Children's Fund. Lilongwe office of the United Nations Development Programme, Malawi. Ministry of Women and Children's Affairs and Community Services, Lilongwe, Malawi.

Whiteside, M. (2003) Enhancing the role of informal maize imports in Malawi food security.

Wiggins, S. (2001) Spatial dimensions of rural development. Department of Agricultural and Food Economics, University of Reading, Reading, UK.

Conclusion

SARAH LEVY[1]

This book has drawn together authors with a wide range of experience and expertise and varying levels of involvement with Starter Pack. They have each examined part of the question which concerns us: does Malawi's Starter Pack programme represent a strategy – or an important component of a broader strategy – for fighting hunger in developing countries? The general answer to this question is 'yes'. Starter Pack has proven to be successful as a way of dealing with chronic food insecurity in Malawi, and lessons can be learnt from the Malawi experience that are relevant to other developing countries.

The book has attempted to present the evidence as objectively as possible. In the words of the famous English Civil War leader, Oliver Cromwell, it has tried to paint a portrait of the programme 'warts and all'. The authors have analysed both the successes of Starter Pack and its problems and challenges. The issues are complex ones, and this conclusion does not attempt to summarize the individual contributions. Instead, it aims to answer some key questions about Starter Pack with a view to helping those who are considering implementing similar programmes:

- In what conditions – agronomic, socio-economic and political – is Starter Pack appropriate and likely to have a positive impact?
- Is the programme adaptable to different contexts?
- What are the potential problems of the programme, and how can they be avoided?
- Could the Malawi version of the programme be improved upon?
- How does the programme fit into the overall policy framework?

In What Conditions is Starter Pack Appropriate?

So far, the Starter Pack programme has only been implemented in Malawi. Would it be appropriate for other countries, particularly for other countries in Africa? The first part of the answer to this question depends on the conditions existing in these countries. On the basis of the Malawi experience, we can reach some conclusions about the sort of context in which the programme is relevant.

First, it is clearly suited to countries whose main staple food is a crop like maize, which thrives on high-quality inputs, but where soil fertility is poor, resulting in low yields and chronic underproduction of food. These are likely to be areas where population density is high and the land is intensively cultivated, as in most of Malawi. These were the conditions for which Starter Pack was originally designed (see Chapter 1). However, even in conditions of poor soil fertility and low yields, it might be argued that free inputs distributions are unnecessary if farmers can afford to buy improved seed and fertilizer. On the other hand, where poverty is serious and widespread, so farmers cannot afford to buy the inputs they need to boost yields, a programme that provides free inputs is likely to be appropriate and effective. In such cases, it does not 'crowd out' private suppliers of agricultural inputs because demand is weak in the absence of the programme (see Chapter 10).

The conditions faced by smallholder farmers in Malawi amount to a serious 'input constraint'. In other words, production of food is constrained less by lack of land or labour – although these may also be in short supply – than by inability to access essential inputs.[2] This situation has been exacerbated in recent years by rising fertilizer prices because of economic liberalization and depreciation of the Malawi Kwacha (see Introduction). It is not unique to Malawi: small farmers in many other developing countries face increasing difficulties in buying the inputs they need, as the terms of trade turn against them following economic reform and markets become increasingly unstable and uncertain.

Second, Starter Pack is particularly relevant for countries where producing food locally is cheaper than importing it. These include landlocked countries where the import parity price of maize is increased by the cost of transporting it over long distances. Chapter 15 points out that in Malawi the direct cost of importing enough maize to ensure national food security – in the form of either commercial imports from South Africa or food aid – is between three and a half and five times the cost of growing the maize at home using Starter Pack. Moreover, taking a 'reactive' rather than a 'pre-emptive' approach (i.e. importing maize when crisis looms rather than producing enough to prevent a crisis) implies serious indirect costs in terms of macro-economic instability and cuts in development programmes.

When assessing whether it is cheaper to produce food locally or to import it, it is worth noting that countries are not homogeneous. For instance, in southern Malawi bordering Mozambique – where maize can be produced at relatively low cost – it may be sensible to rely on imports for part of the food requirement (see Chapter 18). However, this does depend to some extent on relative exchange rate movements, as rapid depreciation of the Malawi Kwacha constitutes a disincentive to importing food.

Finally, Starter Pack is most appropriate for countries with strong – or at least moderately strong – institutional and political development. This is because it is important that the programme be institutionalized within government to ensure continuity and a process of lesson-learning and improvement. At the same time, it should have cross-party political support to minimize the temptation to use it for electoral purposes. It is also vital that there should be local capacity for managing a large-scale inputs distribution, although this need not exist within the public sector: it is best to keep programme management separate under a semi-autonomous Logistics Unit (see Chapters 2 and 3).

Adaptability – the Two Variants

The next question is: how well does Starter Pack adapt to different contexts? To answer this question, it may be helpful to point out that two variants of the programme appear in this book: the original 'Starter' Pack concept and the Malawi version of the programme.

The original design team saw the packs of free inputs with 'Best Bet' technology as capable of jump-starting growth in the agriculture sector (see Chapter 1). The team was working at a time when markets had only recently been liberalized in Malawi, and prices were still relatively stable. Therefore, it was reasonable to expect that farmers would adopt the 'Best Bet' technology to increase commercial production of maize. Their maize sales would enable them to purchase inputs for the next season from private suppliers, and the system would become sustainable within a few years (see Chapter 8). The country would become food secure as marketed maize supply increased.

This vision of commercial maize production leading to agricultural growth as well as food security did not materialize in the unstable market conditions of post-liberalization Malawi. However, it may be relevant to other countries or even to Malawi in a different policy environment at some point in the future. Chapter 19 argues for a return to a more interventionist approach to rural development, including price stabilization and subsidized credit. In such conditions, the original 'Starter' Pack idea might still be viable – although in the opinion of the authors of Chapter 19, it is not the most appropriate food security intervention.

In Malawi, the 'Starter' Pack concept turned out to be a misnomer. The post-liberalization years of the late 1990s and early 2000s, when the programme was implemented, saw serious price instability, which sharply increased the risks associated with commercial farming. This was a bad time to try to promote commercial maize production. The prevailing mood among smallholders – responsible for some 90% of Malawi's maize output – was one of distancing themselves from markets as far as possible. Markets were seen as unreliable, and growing one's own food – always an important part of Malawian farmers' value systems – became increasingly important as a strategy to avoid extreme poverty and hunger (see Chapter 9). Less than 15% of smallholders were selling any maize[3] in the early 2000s, although almost all smallholders grow maize.

In spite of this context – or because of it, depending on your point of view – Starter Pack turned out to be highly effective as a national food security intervention. The evaluation programme found that most rural households in the early 2000s had a maize deficit and were buying more maize than they were selling. Therefore, the key to avoiding hunger was to keep the price of maize – and substitute foods – low throughout the year, avoiding sharp price increases in the 'hungry period' (the 3 or 4 months before the maize harvest). A 'universal' Starter Pack programme produces between 280,000 and 420,000 t of additional maize at national level. This reduces demand-side pressure in the maize market, dampening prices (see Chapter 8).

Thus, Starter Pack is an effective way of achieving national food security, although the mechanism has more to do with demand in the Malawi context than with marketed supplies of maize. The programme also contributes to social protection by reducing the risk of food crisis and strengthening traditional support systems, including local availability of *ganyu* work opportunities.

The fact that Starter Pack was capable of making a major contribution to Malawi's national food security in a very different economic environment from that for which it was designed is testimony to its adaptability. Nevertheless, in other countries it would be wise to proceed with caution, testing and learning from experience before committing large amounts of resources.

Avoiding the Problems

What were the main problems of the Starter Pack programme? It is important to identify the problems with a view to avoiding them in future. Three main areas of difficulty were identified by the Monitoring and Evaluation (M&E) programme: logistics, targeting and extension messages.

With a large-scale programme that aims to deliver packs to nearly three million small farmers in every corner of a country like Malawi – including the remotest villages – it is hardly surprising to find logistical challenges. It took some 5 years for the Logistics Unit to perfect its system for registering beneficiaries and distributing packs of free inputs. In the process, a number of difficulties were encountered and most of them were overcome. Chapter 3 provides an account of how the system evolved, including the problems and their solutions. This is essential reading for anyone contemplating implementing a free inputs distribution programme. Some key points are worth highlighting here. First, rapid delivery of the packs is crucial, as they must be received by the farmers in time for planting. Therefore, each part of the system and its management must be highly efficient. After experimenting with less optimal alternatives, the programme found that the best approach is to subcontract the tasks of input procurement, packing, internal transport and distribution to the private sector through competitive tendering, with a semi-autonomous Logistics Unit (within the public sector, but independently managed) providing overall coordination. The systems used at each stage of the process need to be carefully designed and controls need to be built in. It is also essential that the managers can count on high-level support within the Ministry of Agriculture (MoA) – and government generally – to overcome bottlenecks. If any part of the system fails, there is likely to be late delivery of packs; in the 2000/01 season, packs were delivered too late to be useful in many parts of Malawi.

A second problem area for the Starter Pack programme was when it attempted to replace 'broad' poverty targeting (providing packs to all smallholders, but not to middle-sized or commercial farmers) with 'narrow' community poverty targeting (focusing on the poorest and most vulnerable smallholders) under the Targeted Inputs Programme (TIP). This was unsuccessful and caused serious resentment and social tension in rural communities (see Chapter 11). An important lesson here is that we need to understand more about communities' attitudes to beneficiary selection and poverty targeting before we experiment with it on a large scale. In the case of Malawi, some innovative research into beneficiary selection and the 'right level of targeting' was carried out as part of the TIP M&E programme. It concluded that narrow community poverty targeting is not a good option for large-scale free inputs programmes, although it may be suitable for other types of intervention that aim to target vulnerable individuals.

A third problem encountered by Starter Pack involved the extension system and the messages that it attempted to convey to farmers. The original concept of Starter Pack emphasized the role of the extension system in helping farmers to understand the 'Best Bet' technology. It assumed that there would be strong extension support to programme beneficiaries, so that they could learn optimal methods for planting the seed and applying the fertilizer. However, the evaluation teams found that the technical messages conveyed to farmers via leaflets, the radio and extension workers had little if any impact on them. The messages were confused and sometimes contradictory (see Chapter 12). In particular, there was a conflict between the message based on the 'Best Bet' technologies which underpinned Starter Pack, and the Sasakawa Global 2000 instructions for planting maize (see Chapter 13), popular in some parts of the MoA. The extension approach was also based on the outdated 'transfer of technology' concept and suffered from a breakdown of extension infrastructure (see Chapter 13).

Efforts were made to improve the design of the leaflets that provided instructions on use of the free inputs and to coordinate different communication channels, but extension remained an area of weakness of the programme and farmers continued to use the inputs in less than optimal ways, particularly with regard to fertilizer application. With public funding for agricultural extension in many developing countries under threat following structural adjustment reforms, this is likely to continue to be a problem area for Starter Pack initiatives. Conflicts about the content of the extension messages are also likely to persist while different technology packages are promoted by donors. However, more might be done at the highest levels of agriculture ministries to ensure that farmers receive consistent messages.

Can We Improve on Malawi's Starter Pack Programme?

This book has pointed to a number of areas where Starter Pack, as implemented in Malawi between 1998/99 and 2003/04, could be improved. Before examining them, however, it is worth asking a question about general approach. Should we build on past experience and use the evidence to adjust programmes in the right direction – improving and perfecting them – or should we abandon tried and tested interventions in favour of untested ones? The answer would appear to be obvious. But, in practice, policy makers and donors often abandon tried and tested interventions in favour of untested ones. Why does this happen? I believe that there are two main reasons: first, any programme that is tested will be found to have deficiencies, and it is easy to believe that something new will be better until it is tested and found deficient. Second, regimes and individuals change, and incoming politicians and officials – in donor agencies as well as host country governments – like to make a 'clean sweep'. This is unwise. Abandoning the old is less likely to be successful than adjusting what we already know. History shows that continuity of institutions and gradual adjustment can be a foundation for successful development; continual upheaval and policy reversals are unlikely to be successful.

In the case of Malawi, abandoning Starter Pack in favour of untested alternatives also implies a major risk: if the alternatives do not work, the country will face another food crisis. Nevertheless, the government of Bingu wa Mutharika, who was elected president in May 2004, has taken this risk. In the 2004/05 season, it moved away from Starter Pack towards a different sort of free inputs programme, which had serious design flaws, was implemented late, cost almost twice as much as Starter Pack and did not have an independent M&E component (see Introduction). Meanwhile, the donor community continues to discuss various interventions at the individual/household food security level – including safety nets, home food production and livelihood diversification – without addressing the question of national, aggregate food security (see Introduction).

To return to the question: if Malawi or other countries decide to implement Starter Pack programmes in future, how could they improve on the Malawi experience of the 1998/99 to 2003/04 period? First, I will highlight some recommendations for improvement with regard to increased diversification and sustainable agriculture, forward planning, linkages with development programmes, and the voucher control system. Then I will point to one area – the involvement of retailers – where changes have been suggested, but the evidence is against them in the Malawi case.

Diversification and sustainable agriculture

A key concern about Starter Pack is that it places too much emphasis on maize production. The packs distributed in Malawi included legumes (groundnuts, soybeans, beans or pigeonpeas), but the original plan to distribute different types of cereal and cassava cuttings was abandoned early on. The authors of Chapter 12 argue that 'inclusion of maize and fertilizer in Starter Pack is justified by the need to achieve an immediate increase in the availability of the main staple food', but 'in the longer term, food security and nutrition in Malawi require diversification away from the current dependence on maize'. The programme should encourage farmers to grow a range of food and cash crops, and its logistics capability is now sufficient to deliver different types of seed; the main pack distribution could be accompanied by a system of 'unpacked packs' for crops like sweet potatoes and cassava, which are propagated by cuttings. The advantage of a programme like Starter Pack, which reaches a critical mass of small farmers, is that it can have a strong impact on diversity.

A more controversial issue is whether farmers should receive high-yielding open pollinated variety (OPV) maize seed rather than hybrid maize seed. Opinions on this matter are strongly divided, with the authors of Chapter 12 arguing that OPV can make a positive contribution to sustainable agriculture, while the authors of Chapter 1 emphasize the high yield potential of hybrid seed. Here again, a relevant question is which variant of Starter Pack is being implemented. The original 'Starter' Pack concept expects farmers to be able to purchase seed once they have learned the new technology; if buying seed is not a problem, the optimal choice may well be hybrid. However, at a low-level equilibrium like that of Malawi (in terms of both soil fertility and the prevailing economic environment), OPV may be the more sustainable alternative, even if this is at the expense of yield potential.

Forward planning

A key issue on which there is general agreement about the desirability of improvement is that of the time frame. In Malawi, the Starter Pack programme was implemented on an *ad hoc*, annual basis. The government did not seek to embed it in a broader strategy or to locate it within its Medium Term Expenditure Framework (MTEF). This meant that forward planning was difficult, and donors did not commit resources beyond the short term. However, a medium- to long-term time frame is essential for two reasons. First, it is necessary in order to procure seed of the right type and quality – a 12- to 15-month notice period is needed to produce seed; and second, security that the programme will continue in the medium to long term would permit efforts to maximize its impact on sustainable agriculture, including building better linkages with extension work (see Chapter 12).

If, instead of running six 1-year programmes between 1998/99 and 2003/04, decision makers had decided at the outset to run one 6-year programme, Starter Pack might have transformed Malawian agriculture. This is an opportunity which should not be missed by interventions of this kind in the future.

Linkages with development programmes

An important concern that has been expressed about Starter Pack is that it is not sustainable in the Malawi context. In other words, the government will have to continue to provide packs for beneficiaries indefinitely, with no possibility of 'exit' from the programme. The programme contributes to *alleviating* poverty, but it does not *reduce* poverty, as few of the benefits of food security last beyond the agricultural year.

However, Starter Pack alone cannot be expected to solve all of Malawi's problems. It is an effective way of achieving national food security, but a long-term growth, development and poverty-reduction strategy is still needed (see Chapter 8). Only when reasonable progress has been made in both areas will it be possible to talk of 'exit' from Starter Pack. Nevertheless, it is important to note that without a solid foundation of food security, long-term poverty reduction is much more difficult: with a chronically malnourished population vulnerable to repeated hunger shocks, investments in education, health and other areas of development are largely ineffective. Future free inputs interventions in Malawi and elsewhere could be improved by greater attention to synergies between food security and pro-poor growth and development strategies.

Voucher control system

One area of difficulty faced by those implementing Starter Pack in the early years was that of ensuring that all registered beneficiaries received packs. A system of vouchers was introduced in 2000/01 to help ensure that the packs went to the 'right' people; the vouchers entitled registered beneficiaries to collect a pack from a local distribution centre. However, the evaluation teams found that substantial numbers of packs were misallocated within villages or failed to reach the villages at all, implying 'capture' by people outside the broad target group of smallholder farmers.

In 2003, a distribution control system based on unique serial numbers was introduced to combat this problem (see Chapter 4). This represented a step forward, but further improvements are recommended if policy makers wish to avoid leakages: either the appointment of an independent body to distribute the vouchers instead of local government officials; or a system of spot checks accompanied by financial penalties for those found to be misappropriating vouchers.

The question of retailers

Finally, some high-level policy makers and donors have suggested that it would be an improvement to involve retailers in pack distribution. They are concerned that retailers have been left out of the picture so far, and see vouchers as providing the opportunity for them to take over pack distribution from the private operators currently handling procurement of inputs, warehousing, transportation and distribution.

As part of the research for this book, we commissioned a review of the evidence about retailer involvement in distribution of free inputs. It found that a retail-based delivery system in Malawi would be more expensive than the current system and would mean beneficiaries having to walk much longer distances to collect their packs (see Chapter 4). The current system of delivery lowers costs through economies of scale and competitive tendering to the lowest-price private operators. However, in countries where retail outlets have better coverage of remote rural areas and lower costs, it may be worth considering whether they would be a good distribution channel for the packs.

How Does the Programme Fit into the Policy Framework?

Our final question is about how Starter Pack fits into the broader policy framework. The Introduction to this book pointed out that there are two broad types of intervention to fight hunger: those designed to achieve national, aggregate food security; and those which focus on food security at the household/individual level. Starter Pack – a large-scale programme tackling chronic underproduction of food – is primarily of the first type, although it also makes a modest contribution to food self-sufficiency at household level.

The distinction between these two types of food security intervention is important, and is well understood by Malawi Government policy makers. They believe that universal Starter Pack has a role to play in achieving national food security, while other policy options – such as crop diversification, cash-crop production, livelihood development and safety nets – are important to improve household/individual food security. Unfortunately, there is less clarity within the Malawi donor community. Of late, disagreements between donors about the role of different options have undermined the government's efforts to develop a coherent approach (see Chapter 17).

This impasse provides a reminder about the high cost of not having a clear food security strategy which enjoys support from all the main players. We cannot afford to take our time deciding on a perfect solution to the hunger problem. Not doing anything in a situation of chronic food insecurity like that of Malawi is neither ethical nor cost-effective, as the resulting food crises have a high cost in terms of human misery, the deepening of poverty and the AIDS pandemic (see Chapter 16); and each new food crisis is a further set-back to the poverty-reduction and development process.

So what is the way forward? For Malawi and other countries with similar problems and conditions, I would argue that there is sufficient evidence about the positive impact of Starter Pack to make it a key component of a strategy to fight hunger. Unlike public works programmes, Starter Pack is possible to implement on a large geographical scale. It focuses on increasing food production but – in contrast with a general fertilizer subsidy, which primarily benefits better-off farmers – it is broadly poverty targeted. It costs much less than formal commercial food imports and food aid. It is also cheaper than attempting to provide welfare transfers at the level that would be necessary to avoid hunger in the absence of an efficient national food security intervention (see Chapter 15).

However, Starter Pack is clearly not the *only* element of a comprehensive policy to combat hunger and to secure growth and poverty reduction. In the Malawi context, other elements have been proposed and are being tested, although they still need a process of independent evaluation to assess their strengths and weaknesses. For whatever combination of options are selected, developing country governments need to play a leading role in collecting and reviewing the evidence and feeding it into well-designed strategies capable of convincing international donors. Finally, it is vital that food-security, growth and poverty-reduction policies be embedded in a medium-term policy framework in order to avoid the problem of short-term interventions that are vulnerable to sudden reversal.

Notes

[1] In writing this Conclusion, I would like to express my debt of gratitude to those attending the conference on 'Starter Packs: A Strategy to Fight Hunger in Developing Countries?' at the Rockefeller Foundation's Study and Conference Center, Bellagio (Como), Italy, October 2004. The conference was attended by most of the authors of this book and some special invitees, who provided valuable thoughts, debates and clarifications, which helped me to write this Conclusion. However, the views expressed here are my own and I take full responsibility for them.

[2] Chapter 14 points out that where land and labour are more serious constraints for farmers than inputs – as in the winter season in Malawi – Starter Pack is much less effective.

[3] This percentage comes from the 2001, 2002 and 2003 TIP surveys, and includes all smallholder farmers who sold some maize, even if the amounts sold were small. Those who sold maize were not necessarily *net* sellers of maize: many of them may have bought more than they sold.

Index

Agricultural development *see* Rural development
Agricultural research 4, 15–16, 25, 160, 177
Agri-dealers *see* Private sector *and* Retailers
Aid 205, 236, 243
  Food aid 1, 3, 203, 207, 208, 230, 251, 253, 255, 280

Beneficiaries
  Number of 23, 34, 36–37, 41, 43, 44, 45, 82, 87–88, 89, 149–150, 193, 214
  Registration 32, 37, 42, 44, 45, 46, 50, 55
  Selection 42, 44, 64, 82, 142–144, 146 (incl. note 4), 147, 148, 149–150
Best Bets *see* Technology

Capacity building 73–75, 91, 95, 143, 233, 253, 257, 271, 273
Cash crops 19, 119, 124–125, 137–138, 166, 197, 199, 200, 231, 256, 261, 266, 267, 272
Cassava 17, 103 (note 2), 104, 119, 124–125, 138, 157–158, 161, 170, 199, 203 (note 1), 231, 247
Census *see* Population

Commodity Tracking System 43, 45, 47, 48
Competitive tendering 43, 51, 63–64, 95–96, 242, 283, 287
Constraints 172, 200
  Capacity of institutions 38, 39, 48, 52, 59, 109, 207, 222, 225, 253, 281
  Inputs 8, 17, 22, 25, 104, 123, 132, 134, 160, 164, 196–197, 209, 270 (note 8), 280
  Labour 17, 123, 184, 196, 197, 198, 199, 222, 225, 274
  Land 4, 17, 22, 122–123, 139, 156, 159, 196, 197, 199, 223, 247, 280
Content of packs 4, 6, 23, 38, 41, 43, 160, 161–163, 171, 193, 240
Coping strategies 81, 84–85, 104
Corruption 52, 58, 87, 238, 240, 252
Cost
  Of evaluation 40, 95–99
  Of food aid 5, 208, 210
  Of general fertilizer subsidy 5, 210, 212, 213
  Of maize imports 5, 204, 205, 206, 210, 232, 243
  Of safety nets 5, 211, 212, 214
  Of Starter Pack 5, 23, 30, 42, 49, 50, 211, 213, 237, 242
Cost benefit analysis 5, 203, 236, 243

Cost effectiveness of Starter Pack 5, 29, 32, 40, 49, 213, 215, 280, 288
Coverage
    Of research 78, 80, 81, 83–84
    Scaling down 1, 4, 8, 34, 109, 111, 141, 213, 223, 242, 274
    Scaling up 4, 34, 214
    Of Starter Pack/TIP 4, 23, 36–37, 40, 105, 109, 110, 172, 282, 288
Credit 6, 15, 17, 18, 19, 103, 110, 113, 132–133, 134–135, 208, 270, 271, 272, 275, 281
Cross-border trade 207, 208, 232, 247, 248, 251, 252, 257, 268, 280
    Starter Pack and 255–256

Dambo 193, 194, 197
    Flooding/waterlogging 195, 196, 199
    Residual moisture 195
Decentralization 171, 177, 178, 234
Delivery *see* Distribution
Demand
    For agricultural inputs 24, 118, 124, 129–132, 134, 135, 136, 159, 160, 210, 212, 271, 273, 281, 285
    For maize 107, 108, 159, 207, 208, 209, 210, 211, 248, 249, 251, 257, 282
    Weakness of 130, 132, 133, 134–135, 137, 139, 262, 280
Dependency 32, 33, 118, 121, 240
    Laziness 118, 120, 121
DFID 33, 34, 35, 42, 44, 49, 71, 74, 77, 95, 233, 234, 235, 238, 243
Dimba see Dambo
Distribution
    Control systems 47, 48, 52, 55, 56, 58, 59–61, 64, 283, 286
    Disturbances at distribution centres 55, 56–57, 60
    Of packs 41, 42, 47–48, 60
        Misallocations 56, 57–58, 60, 286
        Timing of delivery 6, 8, 41, 50–51, 105, 185, 189, 193, 194, 213, 282–283
    Of vouchers 42, 45, 46, 59, 60

Diversity/diversification
    Of crops 2, 3, 23, 107, 108, 113, 134 (note 1), 156, 157–158, 159, 160, 170, 172, 184, 198–199, 200, 203, 211, 240, 241, 269, 285
    Of livelihoods 3, 157, 197, 231, 233, 234, 235, 236, 240, 241, 262
Donors
    Attitudes to free inputs 15, 32, 39, 233, 234, 235, 236, 237, 238–241, 243–244
    Ideology 34–35, 39–40, 226, 234, 235, 242
    Relationship with Malawi Government 35, 43, 207, 215, 230, 233, 236, 243, 287
    Response to evidence 39–40, 70, 74, 91, 203 (note 1), 210, 231, 242, 244, 287
    see also European Commission, DFID, IMF, Policies, World Bank, World Food Programme

Early Warning 80–81, 85, 208
Elections (party politics) 6, 29, 30–31, 33, 35, 36–37, 40, 150, 240, 281
Errors
    Of exclusion 143, 145, 148, 149, 150
    Of inclusion 143, 145, 148, 149
European Commission 15, 33, 34, 203 (note 1), 232, 233–234, 236, 239, 243
Evaluation programme
    Comparability 40, 78, 83, 94, 95, 144
    Consultants 70–72, 74, 95–96, 99
    Contracting 73, 95–96, 99
    Design of 57, 69, 70, 77–80, 144, 180
    Dissemination of results 74, 77, 90, 91
    Ethical issues 90–91
    Financial incentives 97–98, 99
    Financial management 95–97, 98–99
    Independence of 6, 9, 40, 74, 77
    Modules/modularization 69, 72–73, 77–80
    Samples/sampling 70, 72, 73, 78, 82, 84, 89
    Training 83, 90, 95

Exchange rate 6–7, 18, 103, 107, 110, 113, 114, 133, 136, 206, 207, 208, 210, 214, 251, 253, 280
Exit from Starter Pack 6, 109, 113, 114, 139, 171, 239, 240, 286
Extension 24, 30, 38–39, 164, 171, 172, 175–178, 285
   Adoption/adaptation of technology 160, 163, 176–177, 189
   Breakdown of 177–178, 186, 189, 222, 240, 283
   Farmer participation 176, 177, 178, 180, 185
   Leaflets 43, 46, 78, 163, 178–179, 180, 181, 182, 183, 184, 185, 186, 187, 189, 283
   Messages 82, 166, 177, 178, 185
      Inconsistency of 39, 163, 165–166, 185, 188, 283, 284
   Radio 16, 178, 181, 183, 186, 188, 189, 226
   Transfer of technology 109, 118, 160, 176, 189, 283
   Workers 170, 177, 178, 183, 184, 185, 186, 188, 189

Facilitation 141, 149, 150
Farm Family Database 37, 43, 44, 52
Farming practices/systems 16, 17, 25, 156, 172, 176, 180, 189
Fertilizer
   Amounts distributed by Starter Pack 38, 43, 46, 130, 160, 161–163, 169, 193
   Application of 38, 172, 186, 195, 283
   Efficient use of 16, 20, 159
   Inorganic 19, 159, 169
   Organic 16, 19–20, 21–22, 159, 169, 184, 196
Financing 204–205, 206, 225
Fiscal position 7, 113, 203, 204, 205, 207, 208, 210, 215, 223, 238, 239, 253, 255, 271
Food crisis *see* Hunger, Crisis
Food insecurity 15, 249
   Chronic 2, 103–104, 111, 204, 208, 214, 215, 220, 229, 230, 234, 244, 279, 280, 286

Indicators of 81, 84–85, 112, 113, 114, 144, 220, 253
Seasonality of 2, 81, 85, 104, 223
Food security
   Household/individual 2, 3, 23, 107, 156, 194, 199, 209, 223, 225, 230, 233, 247, 284, 287
   National 2, 3, 5, 23–24, 103–105, 109, 110, 193, 204, 214, 225, 229, 230, 233, 235, 282, 284, 287
   Strategies 2, 3, 9, 22, 23, 31–32, 39–40, 109, 113, 114, 168, 170, 193, 203, 208, 214, 215, 225, 226, 229–231, 233, 234, 235, 236, 238, 239, 241, 243, 244, 255, 256, 263, 264, 267, 274, 279, 284, 287–288
see also Food insecurity
Forecasting 15, 25, 253–254, 257

Ganyu 17, 24, 82, 104, 107, 112, 119, 123–124, 137, 138, 223, 231, 251, 256, 262, 275, 282
Growth 17, 109, 110, 113, 114, 117, 171, 203, 210, 221, 225, 261, 263, 264, 265, 267, 273, 274, 275, 281, 288

HIV/AIDS 39, 82, 111, 123, 220, 262, 287–288
   Antiretroviral therapy 221, 222
   Food crisis and 223–225, 226
   HIV prevalence 219, 221, 222
   Orphans 221, 223
   Poverty and 220–221, 222
   Prevention messages 226
Hunger
   Crisis 1, 8–9, 15, 34, 80, 81, 111, 193, 203, 207, 213, 223, 234, 236, 242, 284, 287–288
   Hungry period (or season) 85, 104, 107, 110, 157, 170, 193, 207, 223, 282
   Malnutrition 17, 18, 113, 170, 220, 222, 224, 225, 286
   New Variant Famine 222
   see also Food insecurity *and* Food security
Husbandry 24, 123

IMF 204, 205 (note 2), 207, 233, 234, 236, 238, 239
Implementation *see* Management, Of free inputs programmes
Imports
   Of fertilizer 7, 63, 103, 110, 113, 133–134
   Of food 2, 3, 85, 107, 229, 230, 232, 253, 254
   Of maize 18, 204–205, 206, 207, 210, 215, 225, 235–236, 243, 247–248, 251, 253, 256, 268, 280
Income inequality 220, 221
Incomes of small farmers 3, 19, 23, 25, 38, 86, 107, 124, 134, 137–138, 139, 157, 166, 170, 203, 207, 231, 244, 251, 264, 266, 267, 268, 270, 274
Inflation *see* Prices, Inflation
Intercropping 17, 21, 156, 159, 160, 166, 170, 171, 180, 186

Legumes 38, 138, 166–168, 169, 171, 172, 178, 199, 231, 241, 247, 285
   Consumption/nutrition 160, 166, 168
   Quality of 161, 168
   And soil fertility 16, 19, 20–21, 160, 166
   Type of 43, 160, 161, 166, 168
Liberalization 6, 17, 18, 30, 103, 107, 109, 110, 117, 132–133, 159, 208, 209, 230, 236, 238, 280, 281
Literacy 180, 184, 185, 187, 220, 262
Livelihoods 2, 82, 113, 124, 137–138, 156, 171, 220, 222, 223, 230, 235, 261, 262, 266
   In neighbouring countries 247, 248, 251, 255, 256, 257
   see also Diversity/diversification, Of livelihoods
Livestock 4, 138, 157, 198, 266
Long-term view 16–17, 34, 39, 40, 155, 156, 159, 161, 171, 172, 244, 263–264, 275, 285–286, 288

Macro-economy 7, 113, 203, 204, 206, 207, 215, 232, 243, 262, 264, 280

Maize
   Buying of 2, 17, 18, 85, 104, 108, 119, 120, 121, 170, 209, 249, 251, 254, 268, 275, 282
   Farmers' attitudes towards 103, 110, 117–120, 170, 194, 269, 281
   Household deficits of 17, 104, 107, 209, 269, 282
   Improved seed (hybrid and OPV) 15, 25, 38, 103, 105, 118, 137, 157, 160, 161–163, 164–165, 166, 168, 171, 172, 178, 233, 285
   see also Technology, Improved maize varieties
   Production 2, 5, 8, 15, 16, 23, 71, 80–81, 103–105, 106, 193–194, 197, 200, 208, 209, 211, 231, 244, 249, 253, 261, 274, 282
   Incentives 6, 8, 110, 121, 269, 270
   In neighbouring countries 248, 249
   Sale of 109, 110, 113, 118, 119, 124–125, 137, 209, 269, 270, 281, 282
   see also Demand *and* Imports *and* Prices
Management
   Of evaluation programme 9, 77, 93, 95, 99
   Of free inputs programmes 35–36, 40, 42, 50, 51, 53, 58, 171, 281, 282–283
   Logistics Unit 36, 42, 45, 50, 170, 194, 213, 281
Manure *see* Fertilizer, Organic
Mapping
   Community mapping with cards 88–90, 144, 146 (note 4), 149
   Social mapping 87, 88, 144
Markets
   Agricultural input markets 129, 132–134, 135, 240, 241, 242
   Farmers' attitudes towards 120–121, 124, 125, 281
   Imperfections of 8, 130, 132, 133–134, 135, 139, 261, 262, 263, 271, 274
   Labour 262, 264, 266, 270, 274

Index 293

Market development 8, 24, 114, 138, 234, 235, 236, 238, 242, 244, 263, 264, 265, 267, 273, 274, 275
Medium Term Expenditure Framework 32, 213, 215, 226, 285
Methodology *see* Evaluation programme *and* Research methods
Monitoring and evaluation *see* Evaluation programme
Monitoring/policing
   Of free inputs programmes 52, 59, 60, 64, 286
   Of retailers 63, 64

Participatory methods 78, 82–84, 87, 88–90, 93, 144, 156 (note 1), 180
Payments
   To evaluators 77, 96–99
   To implementers 42, 51
Penalties 48, 49, 52, 56, 59, 64, 97, 98, 286
Pests 17, 156
Policies
   Of donors 8, 29, 32–33, 34, 193, 203, 233, 234
   Evidence-based 9, 40, 90–91, 288
   Food security policies *see* Food security, Strategies
   Of Malawi Government 22, 29–30, 34, 35, 118, 165, 193, 204, 206–207, 209, 215, 239, 243–244, 284
   Of Ministry of Agriculture 22, 30, 31, 163, 283, 284
Policy debates 32–33, 117–118, 125–126, 229, 243–244, 287
Politicization *see* Elections (party politics)
Population
   Density 159, 247, 256, 280
   Rural, estimates of 1, 37, 87–88, 103
Poverty 4, 25, 143, 147, 148, 219, 220, 280
   Food and 1, 119, 120, 125, 144, 220
   Index 78, 85–86, 143, 144
   Lines 1, 86, 134, 143, 220
   Poverty reduction strategy 219, 225, 226, 234, 236
   Profiles 86, 87, 105, 144, 145
   Reduction 110, 112–113, 114, 117, 125, 137, 155, 171, 199, 203, 225, 234, 263, 264, 274, 286, 287–288
Prices
   Of fertilizer 7, 18, 103, 109–110, 114, 132, 133, 136, 139, 160, 196, 209, 214, 280
   Inflation 18, 63, 96, 107–108, 110, 138, 206, 208, 211, 214, 220, 251, 273
   Of maize 2, 18, 23, 85, 107–109, 110, 113, 114, 121, 206–207, 209, 210, 211, 212, 220, 223, 225, 231, 249, 250, 251, 253, 254, 268, 269, 270, 273, 274, 275, 282
   Price policies 3, 6, 138, 209, 230, 233, 268–270, 272, 275
   Stability/stabilization 17, 107–108, 160, 212, 214, 232, 264, 268–269, 272, 275, 281
Private sector 33, 42, 51, 56, 61, 64–65, 118, 129, 139, 206, 207, 208, 210, 214, 230, 231, 232, 235, 236, 240, 241, 242, 268, 272, 273, 287
   see also Supply, Of agricultural inputs
Procurement of inputs 36, 42, 45, 46, 161, 166, 172, 285
Productivity
   Of labour 219, 222, 225
   Of maize 15, 16, 17, 25, 159, 165, 197, 267, 274, 280
   Maize Productivity Task Force 18, 22, 25, 29, 159
Public expenditure management 212, 213, 214
Public spending 113, 204, 205 (incl. note 2), 206, 212, 225, 238

Quality control
   Of inputs 161, 168
   Of research 9, 72, 73, 80, 90, 91, 94, 98

Research methods
   Integration of participatory methods and statistics 77–78, 80, 82, 83–84, 87–88, 88–90, 144, 180

Research methods *continued*
    Qualitative approaches 70, 72, 81–82, 117, 118–119
        see also Participatory methods
    Representativeness 78, 82, 83, 93
    Standardization 82, 83, 90, 94, 95
    Tools 87, 88–89, 90, 94, 144
        see also Evaluation programme *and* Surveys
Retailers 6, 56, 61–65, 118, 133, 240
    Competitiveness of 63–64, 287
    Geographical coverage 61–63, 64, 287
    see also Monitoring/policing, Of retailers *and* Private sector
Risk 25, 31, 60, 63, 111, 121, 124, 133, 155, 160, 170, 176, 177, 200, 203, 209, 210, 213, 214, 244, 252, 261, 262, 267, 268, 269, 271, 272, 281, 282, 284
Rotation of crops 17, 20–21, 159, 160, 166, 170, 171
Rural development 110, 113, 203, 222, 264, 265–267, 275, 281, 286
    Regional perspective 256–257

Safety nets 2, 3, 33–34, 111, 112, 203, 212, 213, 214, 225, 226, 230, 232, 234, 235, 264, 265, 275
    Direct Welfare Transfers 113, 208, 214, 255, 273
    Public Works Programmes 24, 63, 109, 112, 147, 214, 235, 239–240
    TIP as a safety net 25, 34, 35, 236, 239, 240, 241–242, 243, 244
Sasakawa Global 2000 programme 118, 163, 178, 283
Scale *see* Coverage
Seed
    Availability 51, 132, 134, 136, 137, 156, 157, 161, 164, 165, 171
    Selection/recycling 163, 164, 165–166, 172, 185, 186, 187
    see also Maize *and* Legumes
Self-selection *see* Beneficiaries, Selection
Self-sufficiency 2, 3, 105, 107, 121, 124, 125, 209, 229, 231, 236
Sexual behaviour 219, 221, 223
    Transactional sex 222, 224

Sexually transmitted diseases 219, 224
Shortages
    Of food *see* Food insecurity *and* Hunger
    Of land *see* Constraints, Land
Social protection 111–112, 114, 213, 223, 282
    see also Safety nets
Social values 119–120, 125, 147, 222, 281
Soil fertility 7, 16, 17, 19–20, 104, 121–122, 156, 159, 164, 195, 196, 197, 223, 280, 285
    see also Legumes
Stabilization and structural adjustment policies 6, 142, 233
Strategic Grain Reserve 3, 8, 205, 212, 214, 223, 230, 231, 232, 233, 235–236, 238, 255, 268
Subsidies 113
    Of consumer prices 107, 204, 206–207, 210, 215, 233, 238
    Of input prices 3, 6, 17, 19, 132–133, 208, 209, 233, 236, 270–271, 272, 275–276
        General fertilizer subsidy 6, 110, 209, 210, 212, 213, 238, 271
    see also Prices, Price policies
Supply
    Of agricultural inputs
        By ADMARC 30, 132, 136–137, 139
        By private traders 24, 32, 33, 109, 118, 133, 134, 136–137, 139, 240, 280
    Constraints 133–134, 136–137, 139
    Of maize 107, 209, 210, 248, 249, 253–254, 268, 269, 270, 275, 281, 282
Surveys 1, 56, 57, 70, 78, 79, 80–81, 84, 85, 86, 93, 94, 95, 96, 104, 110, 122, 123, 130, 132, 133, 134, 141, 144–145, 146, 157, 180, 184, 185, 186, 187, 193 (note 1), 194, 197, 198–199, 203 (note 1), 219, 224
Sustainability
    Agricultural 38, 81, 155–156, 159, 165, 170, 171–172, 285
    Fiscal 204, 243

Index

Sweet potatoes 103 (note 2), 104 (note 3), 119, 124–125, 138, 157–158, 170, 196, 199, 203 (note 1), 231, 247

Targeting 4, 55, 111, 193, 211, 273, 274
    Administrative 142, 273
    Broad versus narrow 23, 111, 112, 141, 150–151, 213, 283, 288
    Community 34, 82, 111, 141, 142–144, 151, 273, 283
    Failure of 86, 111, 141, 144–146, 147, 148, 150, 240, 283
    Geographical 36–37, 142, 150–151, 197 (note 3)
    Self-targeting 142, 148, 149, 239, 273
Technology
    Best Bets 4, 16, 22, 23, 25, 159, 178, 281, 283
    Improved maize varieties 5, 16, 19, 25, 164
    Technical options 19, 118, 171, 186, 265
    Transfer of *see* Extension, Transfer of technology
Tobacco 17, 19, 124–125, 199, 210, 213, 223, 231, 236, 261, 271
Transparency 8, 33, 36, 50, 58, 59, 64, 90, 91, 148, 149, 253, 268
Transportation of packs 42, 43, 45, 47
Tuberculosis 221, 224

USAID 32, 33, 232, 233, 234, 235, 236, 239, 240, 242

Vouchers 6, 55–61, 64
    see also Distribution, Control systems *and* Distribution, Of vouchers
Vulnerability 147–148, 151, 223, 224, 230, 283
    see also Poverty

Warehousing (including packing) 42, 43, 45, 47
Weather 105, 254
    Cause of disaster 8, 213
    Drought 17, 213
    Insurance against bad weather 3, 208 (note 7), 230, 276
    Rains/rainfall 8, 20, 105, 177, 194, 195, 198
Winter season 103, 185, 187, 193, 194, 196, 200
Women 112, 180, 182, 184, 220, 221, 222, 223, 224, 252, 262
World Bank 15, 31, 33, 34, 133, 177, 233, 236, 238, 239, 240, 243 (note 9)
World Food Programme 208, 214, 239 (note 6), 251

CD Included with this Book (see Inside Back Cover)

The Monitoring and Evaluation Archive of the Starter Pack and Targeted Inputs Programmes

Malawi 1999–2003

produced by

Statistical Services Centre, The University of Reading, UK

This archive allows further exploration of the themes addressed in the book. It presents the information generated during the Monitoring and Evaluation (M&E) of Malawi's Starter Pack and Targeted Inputs Programmes between the 1999/2000 and 2002/03 agricultural seasons. The information comes from sources ranging from surveys to participatory research and case studies. Most of it relates to the main season free inputs programmes, but there are also two studies which look at the 2003 winter season programme.

The archive contains the final reports from each M&E study carried out between the 1999/2000 and 2002/03 seasons, as well as annual main reports, policy briefing papers and workshop presentations. A full description of the methodology for each study is included, together with copies of all the instruments used for data collection. The original data gathered by the studies is also on the CD; it ranges from databases to photographic records of participatory discussions and documentation of their outcomes.

In addition, the CD includes:

- annual reports on the implementation of Malawi's free inputs programmes, prepared by the Logistics Unit;
- background papers written in 1998 before the first Starter Pack;
- the evaluation report for the 1998/99 Starter Pack programme;
- two methodology papers examining an approach that was developed by the 1999–2003 M&E programme to generate statistics using participatory methods; and
- reports from a pilot study looking at welfare transfers, which was carried out by Concern Universal (a Non-governmental Organization) between 2001 and 2002, with support from the Targeted Inputs Programme M&E managers.